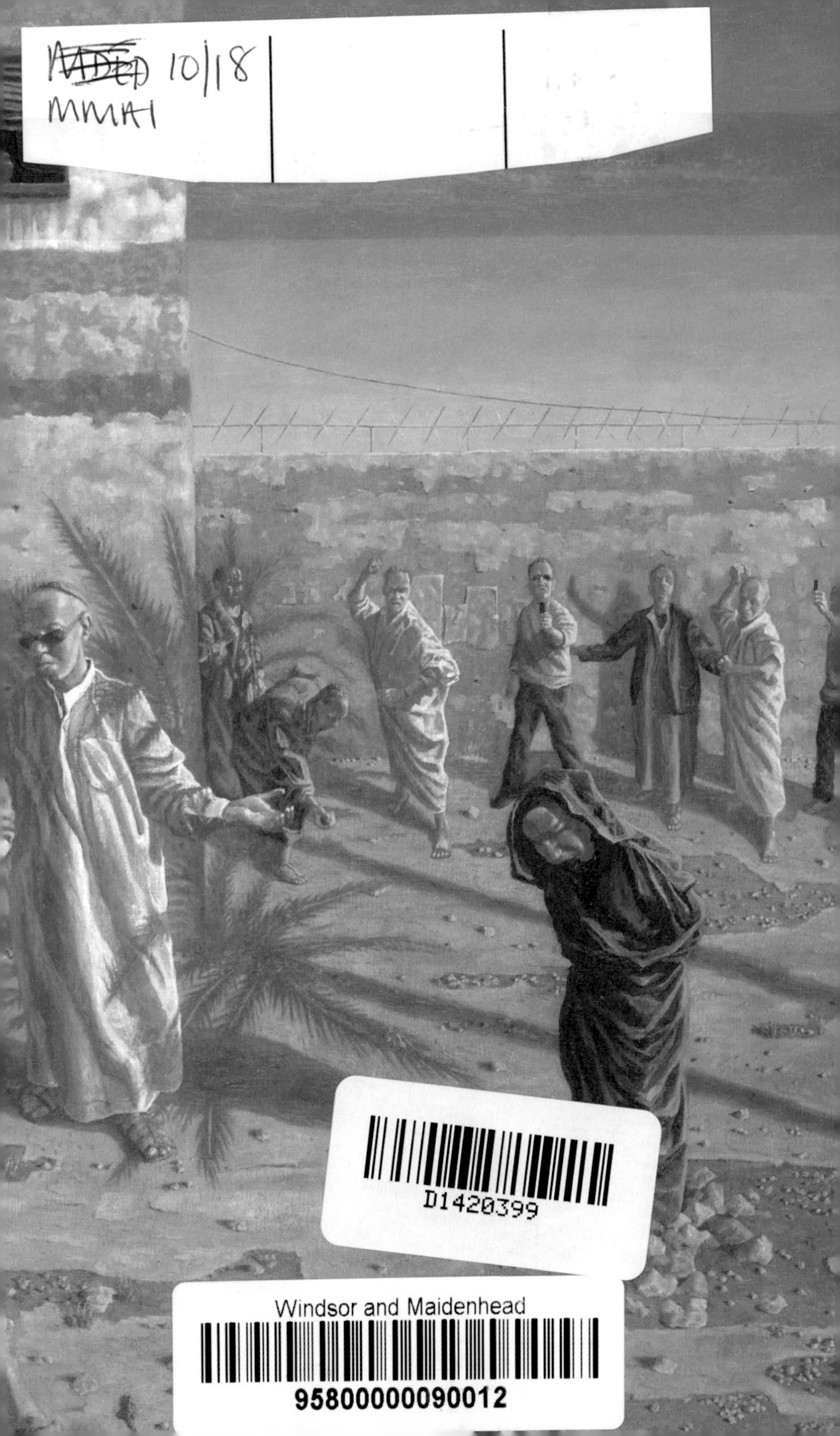

FURTHER PRAISE FOR *IT KEEPS ME SEEKING*

'Of the four themes on which this book is based, two strike a particular chord with an experimental particle physicist like me: "uncertainty is OK", and "we are allowed to open up the window that the natural world offers us". Uncertainly is more than OK, I would say, it is essential to scientific progress, and it's refreshing to see this reflected in what is essentially an examination of humankind's relationship with a Christian god. It is also refreshing to see that the relationship described here by Briggs, Halvorson and Steane sees no inconsistency between scientific exploration of the natural world—opening the window on nature—and faith. Too often have the two been at odds. Humans have been struggling for centuries to find meaning on both the physical and the spiritual planes, and they will continue to do so for as long as there are humans. This book is a rallying call for those quests to join hands. It's a call that will not be taken up by all, but I suspect that those who pick up the dialogue will reap the greatest rewards. After all, the natural world is a thing of great, perhaps spiritual beauty, and the more that science reveals the deeper that beauty becomes.'

Rolf Heuer, President of the German Physical Society
and President of the SESAME Council,
Director General of CERN 2009–2015

'This book crafts a compelling picture of how the desire to know God more closely leads us into all aspects of human life. It takes the reader through the deep, mysterious fields of science and history, using those as springboards for exploring our relationship with God. It is a truly beautiful invitation to look at how science and faith can work together to further our desire for God, which fully showcases the astounding magnificence of God as Creator.'

Justin Welby, Archbishop of Canterbury

From left to right, Andrew Steane, Hans Halvorson, and Andrew Briggs, talking together about the topics of *It Keeps Me Seeking* in the roof garden of St Anne's College, Oxford. Green Templeton College is in the background. (After the photograph was taken, a colleague pointed out a similarity with the poses of Aristotle, Ptolemy, and Copernicus in title page of Galileo's *A Dialogue Concerning the Two Chief World Systems*. Any such coincidence is entirely unintentional!)

FURTHER PRAISE FOR *IT KEEPS ME SEEKING*

'Of the four themes on which this book is based, two strike a particular chord with an experimental particle physicist like me: "uncertainty is OK", and "we are allowed to open up the window that the natural world offers us". Uncertainly is more than OK, I would say, it is essential to scientific progress, and it's refreshing to see this reflected in what is essentially an examination of humankind's relationship with a Christian god. It is also refreshing to see that the relationship described here by Briggs, Halvorson and Steane sees no inconsistency between scientific exploration of the natural world—opening the window on nature— and faith. Too often have the two been at odds. Humans have been struggling for centuries to find meaning on both the physical and the spiritual planes, and they will continue to do so for as long as there are humans. This book is a rallying call for those quests to join hands. It's a call that will not be taken up by all, but I suspect that those who pick up the dialogue will reap the greatest rewards. After all, the natural world is a thing of great, perhaps spiritual beauty, and the more that science reveals the deeper that beauty becomes.'

Rolf Heuer, President of the German Physical Society
and President of the SESAME Council,
Director General of CERN 2009–2015

'This book crafts a compelling picture of how the desire to know God more closely leads us into all aspects of human life. It takes the reader through the deep, mysterious fields of science and history, using those as springboards for exploring our relationship with God. It is a truly beautiful invitation to look at how science and faith can work together to further our desire for God, which fully showcases the astounding magnificence of God as Creator.'

Justin Welby, Archbishop of Canterbury

From left to right, Andrew Steane, Hans Halvorson, and Andrew Briggs, talking together about the topics of *It Keeps Me Seeking* in the roof garden of St Anne's College, Oxford. Green Templeton College is in the background. (After the photograph was taken, a colleague pointed out a similarity with the poses of Aristotle, Ptolemy, and Copernicus in title page of Galileo's *A Dialogue Concerning the Two Chief World Systems*. Any such coincidence is entirely unintentional!)

IT KEEPS ME SEEKING

The Invitation from Science,
Philosophy, and Religion

Andrew Briggs

Professor of Nanomaterials, University of Oxford, UK

Hans Halvorson

Stuart Professor of Philosophy, Princeton University, USA

Andrew Steane

Professor of Physics, University of Oxford, UK

OXFORD
UNIVERSITY PRESS

OXFORD
UNIVERSITY PRESS

Great Clarendon Street, Oxford, OX2 6DP,
United Kingdom

Oxford University Press is a department of the University of Oxford.
It furthers the University's objective of excellence in research, scholarship,
and education by publishing worldwide. Oxford is a registered trade mark of
Oxford University Press in the UK and in certain other countries

First Edition published in 2018

Impression: 1

Published in the United States of America by Oxford University Press
198 Madison Avenue, New York, NY 10016, United States of America

British Library Cataloguing in Publication Data

Data available

Library of Congress Control Number: 2018935365

ISBN 978–0–19–880828–2

Printed and bound by
CPI Group (UK) Ltd, Croydon, CR0 4YY

Contents

1

Introduction

Christianity, at its best, is very atheistic. It is atheistic about all the ill-conceived ways of thinking about God. God as a sort of grandfather figure, or power being, lurking just out of sight, but enormous. Enormous, super-powerful, super-loving, but not actually doing anything very much in the context of AIDS or Ebola or the million shocks that flesh is heir to. Such naïve and incredible notions, verging on the infantile, are what you will hear in many a muddled or woefully thoughtless sermon or song lyric. But that is not the way to know God that Jesus of Nazareth was showing us.

Physical reality has a foundation. There is that which is. This desk that I lean on might not have existed, but it does exist. We live on, or in that foundation. Our bodies are threaded with it; this is utterly inescapable. Deeper and richer concepts such as love, trust, commitment, and perseverance are also ways in which we express the reality of who we are and of what is.

The physical embodiment of those richer concepts is what our communal life together is meant to achieve. That is what Christians think, and atheists and others can largely agree. What Christianity brings is not a new ethic but a new power to do it. The world-wide Christian movement is the imperfect channel of a sort of wellspring that emanates inside us. We find that if we look in the direction that Jesus of Nazareth was pointing, and join the movement that he inaugurated, then our humanity can be expressed in the most authentic and full way, and we receive back love, through a sense of increased togetherness, and through a sense of honest dealing, forgiveness, and a challenge to live by the highest possible standards.

We followers of Jesus make sense of our experience of renewal as follows. We admit we don't altogether grasp the truth of things, because reality extends beyond what we can fully grasp. But we think it makes sense to say that *the foundational reality that is the root of physical existence is also the root or source of the richer realities that human life can embody: love, trust, commitment, perseverance, and so on.*

We find that we are not just forced into being but called into being. That is, we—all humans—and the other animals too, to a more limited extent, are not just forced into existence by the inexorable and blind processes of the physical world. We are also called into the fullness, depth, and height that is accessible to our life. We are called, as people, by one who so calls. We are talked into talking, loved into loving, and forgiven into forgiving.

This book is a joint project between two scientists and a philosopher of science, addressing the area commonly called 'science and religion' from a Christian perspective. This is not a work of apologetics. That is, it is not our main aim to present arguments to defend certain religious propositions. Rather, we would like to show how science both enriches and is enriched by Christian faith. And the first thing we want to assert is that there are not two things here, but two parts of one thing. The one thing is the integrated unity of an emancipated human identity. Emancipated human identity takes science seriously and uses it to the full, and also enjoys the liberty of framing that identity in phrases such as 'a child of truth' and 'a child of love' and 'a child of beauty' and 'a child of mercy' and 'a child of fairness' and, gathering up the colours of this rainbow into one unsettling word, 'a child of God'.

We the authors are not quite certain how to define words such as 'religion' or 'religious', so we decided to approach that in an autobiographical way. We are not particularly bothered whether our experiences and responses are called religious; we are concerned simply that they be truthful and hopeful and intelligent and not self-seeking. It seemed truthful, hopeful, and intelligent to us to open ourselves to the notion that all humans are called upon and valued, and thus encouraged to play a part in making the world better. We think we are called upon and valued by that absolute source and dependable goodness that our teacher Jesus of Nazareth spoke of as 'father'. We understand that such language is a pointer to a mode of relating, rather than a label for an entity, and therefore the father we speak of does not fall under the category of things that can be analysed by passive study. That is why we can't approach this as if we were peering down a microscope, or discussing an abstraction in a philosophical treatise.

Our aim is to improve the general level of public discussion in this area, and to help people with an enquiring disposition to find their way forward. We aim to do this in a way that does justice both to the care that is needed in reasonable thought and to the full breadth of human experience and value.

1.1 Four themes

Four themes run through the book. They are implicit in almost everything we write. It will be helpful to set them out explicitly at the beginning.

1. God is a being to be known, not a hypothesis to be tested.
2. We set a high bar on what constitutes good argument.
3. Uncertainty is OK.
4. We are allowed to open up the window that the natural world offers us.

First, that God is a being to be known. We can't explain what we mean by this, except by inviting the reader into the book and into a larger ongoing journey in which our writing efforts will, in the end, be of small importance. This first phrase is itself hard to get right; the words 'a being' are liable to be misunderstood; please place the emphasis not on those words but on what they introduce: 'to be known.' That is the heart of it. This theme introduces how when we try to enact our deepest duty (and thus discover our fullest liberation), we must do it the right way, because it can only be done in the right way. It is not a path of abstractions, but a meeting of persons. That is why we speak of One 'to be known', but our language struggles. The approach to God is not an approach to another item in a list of items, nor to another person in a list of persons. We simply address ourselves, as directly and honestly as we can, to that which makes our own being possible and calls us onward. One might say not so much 'a being' as, simply, 'Being'—but neither way of speaking captures the situation very well. The only adequate language is the language of the way we live, the way we enact the meeting. This meeting will not resolve for us any item of scientific enquiry, such as the mass of the Sun, or the causes of extreme weather, or the structure of the Higgs field, or the development of the bacterial flagella motor, but it will encourage us on our journey of scientific enquiry, and hold us to high standards of scientific integrity.

Which brings us to our next theme. We set a high bar on what constitutes good argument; in fact, we set the bar as high as we can put it.

If anyone wishes to persuade others by way of arguments in the form of axioms, reasoning, and deductions, then by all means let them do so, but let's not waste time on reasoning which is bad reasoning, or on deductions which don't follow. If a friend tells us that our argument does not work, then let's listen carefully: maybe they are right. And if they are right, then we must throw the supposed argument away. It is not an argument. To bring it out again would be like offering a stone to one who needed bread—an exercise that is not only useless but also hard-hearted.

Often we are left without sure-fire arguments. We can only do our best and say that what we have come to consider is, in all honesty, well grounded, but there remain many doubts and uncertainties, and many things we just don't know. So be it. This is what it is to be alive. We want to affirm that such uncertainty is OK. This is our third theme.

The theme of uncertainty is central in scientific endeavour, and familiar to all good scientists. One of the great strengths of science is that it trains us to consider not only what we think we measured, but also to pay attention to all the uncertainties surrounding what we have been doing. In order to make progress in science, we need some initial idea of what we suspect may be the case, but we also need a bit of flexibility, the awareness that we might be wrong, and we must not overestimate (to ourselves or others) the reliability of our deductions.

In the context of relationships between persons, on the other hand, the issue of uncertainty is much harder to manage. Consider, for example, the situation of Othello, the Moor of Venice, so painfully and inexorably drawn out for us in Shakespeare's play. Othello is presented with evidence of his wife's infidelity, not abruptly, but gradually and insidiously. The evidence has been fabricated, but he does not know this. What is he to do?

This example illustrates how uncertainty in human life in general is not so easy to manage as it is in the context of dry data and impersonal tests in a scientific laboratory. Christian communities, recognizing this, have tried to encourage people not to give too much anxious attention to their uncertainties about God, but often they have ended up creating an oppressive atmosphere in which people cannot readily admit that there is no such thing as certainty. In relationships of persons, there is no certainty, there is only the decision to trust or not to trust, and to keep seeking. We want to affirm that uncertainty about many of the issues we face is OK. It is OK to be unsure about exactly what it means to have 'treasure in heaven', for example, as long as we keep in mind that it is about being generous and not bothered about fame and worldly recognition. It is OK to be unsure whether the Gospel writers occasionally have made mistakes in their report, as long as we give them credit for overall honesty and a fair level of general competence. It is OK to be uncertain exactly what we mean by 'God', as long as we keep hold of the notion of high moral stature, and we don't arrogantly dismiss teaching that is offered by long-tested sources of wisdom, just because we can't quickly settle it comfortably in our heads.

Human beings are not tracked vehicles like tanks and bulldozers, creeping forward in safety and pummelling the ground; we are light-legged walkers, whose every step involves both faith and uncertainty. Our foot hovers over we know not what, nor whether it will hold, before we place it down. And that is just one step.

And now for our fourth theme: we are allowed to open the window that the natural world offers us. In this theme we want to assert a sort of opening of the windows of the house, a release of the human spirit, a reaching up, an allowance to be fully human and to give honest expression to all our capacities. In this image, the house is the natural, physical world, the stuff that is made of the quark field and the electron field and the electromagnetic field and so on—all the stuff that physics shows us. This is our house; we are physical. The philosophical position called *naturalism* wishes to add that this is also our home, and the whole of what is. We would like to say that we welcome much of what naturalism affirms, especially about the value of evidence and reason, and about caution over existential issues, and about affirming the value of ordinary physical stuff in the here and now. We welcome all this. But we want to say, dear naturalism, that it is impossible to know for sure that the physical world is the whole of what is, and furthermore there is plenty of reason to suppose that it is not. For the whole world gestures onwards, beyond itself. It gestures towards a love that struggles to be born, a love that it does not show, but could. It gestures towards a love that *we* do not show, but might.

What does that mean? We hardly know. But we share the inheritance of humankind: the sense of injustice about how things currently are, and the unquenchable hope, the sense that our hearts do not reside fully in this physical house, but in a larger home that, together, we are discovering and helping to bring and share. This is highly subtle, of course, and liable to be misunderstood. We don't want to run away to somewhere else, some other-worldly place of inane or smug forgetfulness. We want the peace and justice of our true home to be realized right here in the physical house of life on Earth. That is what genuinely Christian community life is about. And in our fourth theme we want to assert that we are allowed to think this way. We our allowed to throw open the windows and let the warmth shine on our faces and let the fresh air come in. We are allowed to be fully human, spiritually open-ended, allowed to enact our deepest duty and thus discover our fullest liberation.

1.1.1 Opening the first theme

Earlier, we called the word 'God' an unsettling word. It is also an unsettled word. Different people use it to mean different things. They even use it to mean different types of things.

The word 'God' is thought by some to refer to a supernatural entity whose impact in the physical world can be assessed by controlled experiments, in which the experimenter is a passive observer, not a partner in a loving relationship. This is like the way a Nazi scientist such as Joseph Mengele understood the word 'Jew'. This is such an extreme departure from an intelligent and correct use of the word 'God' that we fear the word may be unrecoverable and we will need a new word. However, we have not yet given up the effort to recover the correct usage, and we will try to clarify that in this book.

In the usage of the word 'God' that we wish to recover, the following question is meaningless: 'Who or what made God?' This question is meaningless in a similar way that the question 'What colour is philosophy?' is meaningless. It is a category error. (A *category error* is the error of presenting something or someone as if they were an entirely different type of thing from what they are.)

In the usage of the word 'God' that we wish to recover, the following phrase is self-contradictory: 'the God hypothesis'. It is self-contradictory in a similar way that the phrase 'a square circle' is self-contradictory. It attempts to put God in a class that God cannot be in.

The first theme of the book is an effort to show what is wrong with these misconceived ways of speaking, but this cannot be done by the type of argument which itself adopts that very way of speaking, a way which fails to grasp the very nature of what you are dealing with. We can only say, 'no, not that,' and then try to show another way by exhibiting it in action. But it is useful to get this initial 'no, not that' out in plain view at the start.

Another question that is itself questionable is 'Does God exist?' This is because as long as someone is asking the question, then that which they have in mind under the symbol 'God' is not that which is truly God. This is like asking 'Is reality real'? The word 'real' ultimately corresponds to that which *makes sense*, to that which different people can share their experiences of and agree upon. God is one who *makes sense* of all the threads of the universe, in the widest and deepest way. Consequently, the appropriate usage of the word 'God' is to agree that this word indicates that which makes sense in the fullest possible way, and our job is to discover what that reality is. By 'make sense' here we mean not just the sort of sense which analytical thought might make, but also the deeper insight which includes wisdom and empathy. Science contributes, but it does not contribute everything, because science does not include personal encounter—or, if you want to include personal encounter under the label 'science', then you must undertake that personal encounter when you enact your 'science'.

In the above we have employed the word 'God' as a title, or a signpost towards a certain kind of role. We can also learn more intimate names, such as 'the liberator of the slaves from Egypt' or 'the caller written about in the account of Abraham' or 'the father of whom Jesus spoke' or 'the one whom Michael Faraday recognized,' or 'the reconciler known by Desmond Tutu'. The idea is that the possessor of these more intimate names is also the proper carrier of the title 'God'.

We said at the opening that the book is an attempt to clarify, and now we seem to be writing in a rather subtle way, a way which some readers will suspect of double-speak and slipperiness. We contend that by alerting readers to the utter futility and misleading, misdirecting barrenness of phrases such as 'the God hypothesis', we have already made a helpful contribution.

1.2 Fear and Sleepy Dust

In our fourth theme we assert the human freedom to reach for aspects of reality which are not immediately apparent or self-evident but rather are emerging and gestured towards.

We would like to say here that there is no need to be excessively fearful of this. The atmosphere of modern-day academia, and of large parts of modern life, is to veer nervously away from all types of thinking which might be in danger of 'letting the side down' by suggesting that atheism was a mistake after all. One can understand why people would (and should) be nervous of the ills that are done in the name of religion, but an atmosphere of leaping immediately to suspicion at the least sign of evidence that one may be mistaken is not a healthy atmosphere.

The natural world has mathematical harmony built into it at a deep level. This is itself quite striking and may give one pause for thought. Another thing that seems to be woven into our world is its ability to support and produce deeper and deeper ways of being in community. Communities of bacteria led to and now live alongside communities of fish which led to and now live alongside communities of human beings. Some people say there is no meaning in this great process, no sign or writing in the sand. It is just the way the dice fell. But, as a matter of logic, that conclusion is unwarranted, and there is nothing wrong in admitting this.

When we say such a conclusion is unwarranted as a matter of logic, we are not saying we know for sure that it is false. We are merely saying that it is not forced on us by the data.

We (the authors) think that the emergence of life and death, and love and hate, on planet Earth suggests that the universe is freighted with meaning. It is freighted with meaning in somewhat the same way that a written text is freighted with meaning.

We want to say that it is quite reasonable to accept this.

This rich process is not just a line of tumbling dominoes, nor a whirlwind, but a story, a narrative. It tells a tale. But a weight of meaning in the very fabric of the world is unsettling for human beings. It is even, in some respects, frightening. And fear drives people to strange behaviour. So you will find otherwise highly intelligent people resorting to the strategy of sleepy dust. They will stand before you and, eloquently, and at length, scatter the sleepy dust.

'No sign, no symbol here!' they will say. 'Don't worry! It is just dice and dominoes!'

They will say it more subtly than that, of course. They will say 'Darwin has shown us' or 'science has shown us' that it is all just dice and dominoes—blind chance and necessity. But that observation is beside the point. It is a category error. The fact that evolution is a natural process does not tell you whether it signifies anything as a whole, because the issue of its overall signification is at another level, in a different category.

The argument advanced by Richard Dawkins in this area is logically equivalent to the argument that the film *Toy Story* is not about friendship, because it is just electronics. It is just a sequence of electric currents in a silicon chip and glowing pixels on a screen. You see, the argument from physical process to absence of message is not an argument; it is illogical. Noticing this will, we hope, help to blow away the sleepy dust. It leaves us free to consider, in all seriousness, the message that evolution conveys.

It is not a message of pitiless indifference because we are in it and we are not pitiless. It is a message of growth and improvisation, pain and perseverance, diverse community, and a message of great creativity given to the world, and to us. We will say more about this in Chapters 12 and 13. It follows from this that the natural world should not be regarded as neutral and without value, nor should it be regarded as our property, as if we could do with it as we liked. Rather, since the world is the conveyer and supporter of meaning, we can find in it a meaningful role to play.

1.3 Three example issues

We will now comment on three questions that commonly arise in discussions of science and religion. The questions were posed to one of the authors for a discussion event jointly organized by Christians in Science Birmingham and the University of Birmingham's Atheist Secular and Humanist society, to whom we are indebted.

1.3.1 Does Christianity impede the progress of science?

The answer is: *Well of course not.* It practically invented science. Or at least, the striking progress of science in the modern era had many of its roots in Christian theistic belief, and for four hundred years the Christian community has largely nurtured science and done it well. There have also been examples of nasty narrow-minded opposition to specific projects or ideas or sometimes people, but these are the exception not the rule.

In this reply we reserve the right to use the word 'Christianity' to mean something close to the notion of *learning from Jesus of Nazareth*, as opposed to the power games that have sometimes self-identified under this label. In this we admit the failings of much genuine Christian discipleship, but we also ask for reasonable effort not to allow charlatans to hijack the label 'Christian' in times when it was fashionable or useful for gaining political influence or social advancement.

Following Jesus of Nazareth and recognizing the sort of 'kingdom' or way of life that he talked about includes, as part of it, doing science as well as we can. We talk about loving God with all our heart, all our soul, all our mind, and all our strength. What that means in practice is this. Loving God means *joining in with God's great project.* It means joining in with a project to emancipate humankind and take care of planet Earth and adjudicate wisely among the other species on it. Science is a significant part of this. Science is a welcome, deep, and very important part of the way of life that followers of Jesus call the kingdom of God.

A thorough presentation of the Christian contribution to science will scrutinize both the positive aspects and the tensions as they have arisen. In this book we shall not attempt such a thorough historical analysis. Our comments here merely outline our honest and well-informed conclusion, in response to a question which has been asked.

1.3.2 Is Christianity incompatible with evolution?

Come again? You would like to know whether Christianity—that is, learning from Jesus of Nazareth about how human life is to be lived, and recognizing one's need of free forgiveness, and living from the prime motivation of gratitude and compassion, and seeking to allow the great source of wisdom to mould one's priorities—whether all this is compatible with the biological and geophysical processes of life on Earth? Well it's sort of a crazy question. It's like asking, 'Is learning the violin compatible with the fact that trees grow from seeds?' Or it's like asking, 'Is atheism compatible with evolution?'

The answer is 'Well of course. Of course Christianity is compatible with what happened. *Christianity is a recognition of the nature of what happened.*'

Some Christians have been taught to be nervous of Darwinian evolution, and who can blame them when it is served up the way some people want to serve it.

The historical background is as follows. The main facts about evolution were gaining their scientific credentials during the nineteenth century. Scientifically informed Christians initially shared the same kinds of reservations about the strength of the case that other thoughtful intellectuals articulated. Admittedly, large parts of the Church world-wide were suspicious about Darwin's work, and slow to be convinced. But this is, in part, because of the issue of human value. The scientific analysis was caught up in political and religious struggle from an early stage. The facts about gradual accumulation of change through natural process were not always made available in a fair and balanced way; instead, they were coloured by some as a triumphant vindication of atheism, and then offered to the general public in those colours (i.e. false colours). But the type of theology that is disproved by evolutionary biology is only a rather simple-minded kind that intelligent theological reflection never needed in the first place. (Francis of Assisi would have understood this very quickly!) And as the scientific case strengthened, Christians in mainstream science, along with others, came to accept the Darwinian picture. As decent scientists, they provided some of the arguments, and found some of the evidence. They then puzzled over what this was telling us about human nature, and the divine nature, as did everyone else. And, of course, not all Christians were willing to accept the evidence. But the direction of travel is that they were coming to accept it. The acceptance by the wider culture of the Darwinian account of biology was in fact promoted by the Anglican church in the UK at the end of the nineteenth century. Yes, you heard me correctly. The Anglican church deserves credit for being intellectually serious about this and helping people get a balanced view of it.

The twentieth century then saw two developments which thwarted this process. These were eugenics and biblical literalism. During the early part of the twentieth century, the concept of 'struggle for survival' and 'improving the gene pool' got carried over into various deeply ugly ideas. The idea of sterilizing certain sections of society was talked about, for example, as if it would be the 'scientific' or 'enlightened' thing to do. And, note well, we are not here talking about the elimination of debilitating genetic diseases. We are talking about racial superiority and genocide. So you can see why some people thought there must be something wrong with Darwinian evolution. 'If that is what it means, then we'll stick to our miracle stories, thank you very much.' Of course, all this was an abuse of science. Our point is that when science is abused in this way, you can understand why people might become suspicious of it.

The other development, less immediately horrible, but presenting long-lasting problems in its own way, has been the notion that authentic Christian commitment requires a completely, or almost completely, literal reading of the account of the Garden of Eden in the Bible. This is an incorrect handling of the Bible, but it has been taught in many churches, in part in reaction to overly dismissive attitudes to biblical texts that can be found in other quarters.

The situation with publicly funded education in America remains unresolved. Surveys show that large numbers of people reject the standard scientific account of biological history, but the subtext here is not primarily about science as such. The problem is that there is a widely believed narrative along the lines that evolutionary biology goes with atheism, so if you don't want the latter, then it follows that you must loyally reject the former. Words like 'evolution' and 'Darwinian' have become trigger words, or code for a whole world view. Education will be promoted when this issue is acknowledged and a more neutral way to present the science is agreed upon.

In recent decades there has been a renewed effort from spokespeople for atheism, trying to promote the idea that natural selection and the rest of Darwinian evolution somehow belongs to atheism. It doesn't. Scientific results do not belong to any particular religious position; they belong to all. But what religious positions must do is to reckon with science and take it seriously, not just in its ability to show the detail and structure of natural processes, but also in the way this feeds into our sense of what it all amounts to, what the significance is. The significance for us is especially what it may tell us about our own role: how human life should be lived.

Some commentators choose to summarize the scientific findings by reaching for metaphors such as selfishness, aggression, and blindness, but we don't necessarily have to buy into those metaphors. Randomness in natural processes is also often portrayed as some sort of defect or problem. But randomness could also be named openness. It is freedom from micromanagement. That is, arguably, a good thing. This entails risk. If intrinsically good things are uncontrolled, then they can go wrong: badly, painfully, terribly wrong. But without risk there is no freedom. These brief phrases are not going to solve or make acceptable the unsolvable, unacceptable problem of pain and suffering. Our point here is simply that openness, and the freedom that results, is something we can think about in a balanced way. We don't need to overlay it with misleading adjectives such as 'lurching' and 'meaningless'.

An individual turn of the dice may be meaningless, but the fact that the dice can turn more than one way is highly meaningful. And the fact that the natural world is so configured that these turns can lead to structures of great depth and richness is even more meaningful still.

The randomness, though significant, is not the main picture. Look at any living thing: it is full of structure. Look at the leaves on a tree, the wings on a dragonfly, and, yes, the wings on a malaria-carrying mosquito too. All are full of structure. Think of the marvellous factory that is a single cell. Pattern and structure throughout. It is staring us in the face. All this is because the natural world has profound mathematical harmony built into it at a deep level, as we said earlier. This harmony emerges again and again in biological forms.

In summary, on careful reflection, the mainstream scientific account of biology can be received by the thoughtful follower of Jesus as an important part of our overall education, with many interesting lessons which we will present at greater length in Chapters 12 and 13.

1.3.3 Is Christianity useful for moral and ethical thought in science and technology?

The answer to this is like the answer to the first question. Christianity in the world today is a mixed bag because it is enacted by a mixed bag of people. However, the driving forces at the heart of it are good. That is what we (the authors) think.

We are asked to consider here whether Christianity is useful. Well, if the central ideas of genuine Christianity were not true, then we would not be very interested in them. So the question 'Is it useful?' is not the primary question. The primary question is 'Is it true?'

However, on the whole, things that are true and beautiful often turn out to be useful too. So let's think about this.

When we think about moral and ethical thought, an important point to keep in mind is that there is broad agreement for much of it. Most of the central ideas of morality and ethics have been worked at and talked about in a range of cultural settings, and many of the central ideas are in common. They are a human universal. An atheistic understanding of ethics and good behaviour is mostly in agreement with a Christian understanding, and a Buddhist one, and a Hindu one, and so on.

When we say that Jesus was a great moral teacher, we don't mean he introduced a new morality. There is no such thing. Rather, he affirmed the universal Tao or Way, and made it real and human. So, for example, he pointed to the fact that at its heart is love, non-violence, recognizing the concerns of other people, renouncing revenge, challenging oppressive power-structures by non-violent but determined truth-speaking, and so on. Ideas like this were already mentioned by other people such as Buddha, Kongzi (Confucius), Lau Tzu, Aristotle, Socrates, fellow

Jewish rabbis, and so on, but Jesus of Nazareth changed the emphasis and clarified the priorities.

What he was doing was, first, living by it and, second, showing how even in the context of grief, violence, and being taken advantage of, we can already possess the heart of life, the pearl of great price.

What Jesus got crystal clear was the remarkable idea

The greater context of our lives is for us, not against us.

We are not situated in a battle to wrestle goodness from an oppressive or aloof or capricious or mindless universe. No. We are partners in a work of creation and we can live tremendously meaningful lives in the here and now, right in the middle of our pain. How happy if we grieve and mourn, because it means we have hearts that care about other people. It is the peacemakers who will be called children of God (yes, the peacemakers, not the people who sing in church on Sunday). How happy are those who are unpossessive—the whole world belongs to them. He really means it. If you can get to knowing what that means, you will possess the world in a profoundly fulfilling way that the richest oligarch on the planet may never know.

We think atheist philosophy can agree with much of this.

What is different about the Christian attempt at community life is not necessarily that it arrives at a different ethic, but that it gives a great encouragement to be brave enough to actually live by that ethic. Also, it gains access to help. This help, which we call the Holy Spirit, can liberate in us a fresh power to overcome our faults and failures and, falteringly, start to live better. Still hopelessly stammering and hesitant, but nevertheless, yes, becoming a bit less miserly, a bit more generous, a bit less lazy, a bit more engaged with the project of being human. A bit more putting into practice the theory of wisdom in thought and love in action.

That is why Christianity is useful. It is useful because it will make a difference. It will persuade a businessman to resist the dodgy tax scheme, and it will persuade a loan shark to make a break with his profession, and it will persuade a drug addict that there is a way out of their addiction in a community and a foundational relationship that will not break faith with them.

Now let us come to what we suspect is the issue the questioner really wanted us to grapple with. This question is about what to do in the case of moral dilemmas. Dilemmas such as science applied to weapons, and to abortion, and to genetics, and to prolonging life, and to ending life, and so on. Can Christianity help with this?

The answer is 'yes it can', partly because of the encouragement to live better which we have already mentioned, but also because we have structures to draw on. The world-wide Christian church has put a lot of time and effort into developing centres of learning, universities, legal framework discussions, democratic political discourse, well-informed charities such as CMS, Oxfam, World Vision, Tearfund, and many others. We are not the bunch of religious ignoramuses that some people like to paint us. We are busy developing an AIDS vaccine and quantum computing and earthquake prediction and legal frameworks around fair trade and international diplomacy and microfinance. We are well placed to contribute intelligently to discussions around machine learning, stem cell research, organ transplants, and the like. And our position is not necessarily the simplistic one of merely outlawing questionable behaviour.

We admit that religion generally does seem to attract to it large numbers of somewhat bewildered and bewildering people with muddle-headed notions and objectionable ideas. The staff of Christian churches put in a lot of work towards trying to be friendly to such people while not allowing their muddle-headedness to dictate policy. And we have our own characteristic problems. The Christian community can often seem to be conservative and slow to change, but then suddenly it can produce breathtaking bursts of life.

1.4 What is this book about?

Christian thinking brings a long memory and the view that we must go forward together as a community of interdependent beings. The idea that humans are a bunch of atomic individuals each asserting their rights is a deeply misconceived picture of the human community.

We love the African concept of *ubuntu* or one-another-ness. This plays out in the way churches run toddler groups, and environmental awareness groups, and men's groups, and women's groups, and marriage courses and carol services and food banks and art exhibitions and drop-in centres and alcoholics groups and debt counselling and prison chaplains and so on and so on. It also plays out in deep and long-term thinking about fair trade, environmental policy, north–south divide, good governance, and so on. Christian action often takes place unlabelled as such, and away from church buildings, as people exercise their gifts in their ordinary place of work, or when one person or family goes to live alongside another. Through all this activity, sometimes muddled, but largely good-hearted, the Holy Spirit is alive and kicking and saving the world through a billion acts of kindness and intelligence.

By such means, the ever-gracious Caller-into-being is going to change the world of human hearts, and those human hearts are going to make informed decisions about science and technology, and that technology will shape the cathedral called Earth in which our children will sing their lives.

All this is what our book is about. We want to make poverty history, and we want to make poverty of imagination about God history. We do not want to waste time on bad arguments. And the way in which 'science and religion' is often framed as a sort of awkwardness-management exercise for religious types is uninteresting to us.

Our central theme is that when we approach God, it is with the whole of our personhood, and we do not begin by insulting God with microscopes. Rather, we open ourselves to the notion that God can be trusted and even, yes, *known*, and it is this open-hearted reaching out, this emphatic liberty to seek and to be received, which is largely what we mean by religious life.[1] Science is part of this, not an alternative to it. But it is fundamental to this whole enterprise that we must apply the techniques of science appropriately, and that means enjoying science at its best. Science is about how to discover the structure of the physical world of which we are a part, not a means to reduce ourselves and our fellow human beings to mere objects of scrutiny, and still less a means to attempt the utterly futile exercise of trying to do that to God.

---

[1] A series of questions, structured as a study guide to *It Keeps Me Seeking*, can be found at http://www.oup.co.uk/companion/seeking.

2

A conversation about the themes

We, the three authors, have had many stimulating discussions in writing this book. Occasionally we have been in immediate agreement about a topic, but that was relatively rare. More often we have found ourselves wrestling with concepts, refining our thinking, and sometimes convincing one another to change the way we approach things. To give the reader a flavour of this, we recorded two conversations between the three of us. One was in the Briggs' garden in North Oxford late one sunny afternoon in May 2017, the other via computer link (Skype) between Oxford and Princeton in September 2017, after the text was mostly written. We have transcribed the conversations and edited them only very slightly to make them easier to follow. If the result suggests that these are issues which we are still thinking about, and where our ideas are still evolving, then that is an accurate impression.

The present chapter consists of the first part of the September conversation. We have indicated the contributions by our initials.

2.1 First theme: the approach to God

HH: I want to begin by being a bit of a devil's advocate against things that I was (and remain) very supportive of in our writing of this book. I want to start with the first theme: that God is a being to be known, not a hypothesis to be tested. I'm wondering, in reflection on that, are we not engaged in some sort of doublespeak there? Because, of course, on the one hand there's an obvious sense in which anyone would say God is not a hypothesis. However, there is a hypothesis that is up for questioning: the hypothesis 'God does exist'. Are we, in the way we're stating that theme, trying to detach ourselves from actual commitment in a way that would be more honest to our subject matter? What are we trying to say?

AS: I would say that the question 'Does God exist?', to me, has already gone wrong, because it is like one of those paradoxical statements such as 'this statement is false'. If I hear the question, then I am ready to answer 'yes' in so far as the word 'God' really refers to the One who has that title, but the very fact that the questioner is asking alerts me to the fact that they are not truly referring to God. They are using the same word for some other notion, some sort of something which could possibly not exist. So the whatever-it-is that they are asking about clearly is not God. They are obviously in a muddle. How should I reply then? The best reply I can offer is something like, 'Let's talk further'. Or I might say something more provocative, such as: 'that's like asking, "Are perfect numbers composite?" ' (The fact that you need to ask shows that you don't know what perfect numbers are.) Or I might even say, 'no', because it sounds like what they mean by 'God' in their question will turn out to be all wrong.

HH: In a way we're placing emphasis on the fact that many people have assumed they know exactly what the debate is about. But we're asking people to reconsider the debate from a new point of view.

AS: Yes.

HH: Or reframe the questions. Do we think that somehow by changing that perspective, we have alleviated what many people would consider to be the intellectual stumbling block of undertaking this kind of religious belief? What effect do we hope this will have on our listeners?

AB: I suppose one way of looking at it is to say that if you try to ask, does God exist or not, you could then ask: What sort of evidence would help you in addressing that question? And it's actually quite hard to think of evidence that would help you to address that question, because your choice of evidence is so biased by your own presuppositions. And there remains a basic difficulty which is unavoidable because it is owing to a very good and unavoidable consideration. For the proof of anything requires something more absolute than the thing that you're trying to prove, to which you refer the proof. This is connected to the sort of concerns that Andrew was raising. That kind of deductive approach is not going to work because it would contradict what we might finally understand by God. A more helpful approach may be to say, if we start with the premise that there is God who is there (and you can think of analogies of axioms in Euclidean geometry), then let's see where that takes us. And let's see whether where it takes us makes sense and seems to be good and true. And I would say that that is indeed my experience and that that is the direction of travel that I go in when I start from that point of belief.

HH: Yes. Since each of the three of us has had slightly different conceptions of how we are doing this book, our efforts really come into focus in different portions of the book. There was in one sense no single vision for exactly what this would be. However, it seems to me that a common theme that's not stated as one of our four is that we don't intend to be giving an argument. We don't intend for a sceptical reader to be able to start with page 1 and we give them premises that they accept and then by the conclusion they should have accepted our belief. It seems to me that what we've done is presenting more of a narrative. We're giving a picture, almost a portrait, saying: Well, this is how we see it. We don't expect necessarily for someone to be convinced but here let's point at this part, look it is beautiful, and this part is related to that part and so on. Is that to some extent the way each of you see the genre of how we're presenting ourselves in this book?

AB: I think that's right and I think that, as you say, we all started from slightly different points because we have each of us lived our own lives and travelled our own pathway of thought and life and faith and belief. And so it's as if we've got different descriptions, but then we see that they are different descriptions of the same thing. I think for me that's been part of the pleasure of writing the book together actually.

AS: Yes, I agree that what we're doing is showing an approach. It is an approach that is on show to share with the reader, a way of approaching human experiences, and questions thrown up by science, including a way to knit them together in an intellectually coherent and satisfying way. It is also an approach that we realize we have to check all the time for moral coherence as well. Is it tending to encourage us in ways we sense to be valuable and good? Now if it weren't heading in that direction, we would start hesitating and asking: hang on, what are we really getting into here? So, to repeat, that's what sort of book it is. And I think I'd like to add another thought. I think again and again what I'm also wanting the reader to understand is that they don't already know what we are offering here. I want the reader to feel that they don't already know what it is that we might mean by God, or what it might be like to be one who is in the 'school' of Jesus of Nazareth. I want the reader to not preconceive what that is. Because maybe it is different from what they thought. I want them to experience that sudden feeling of being taken aback and saying: 'Oh, it's not quite what I thought.' That's the other thing that I hope is going on.

2.2 Third theme: uncertainty

HH: This relates to another point I wanted to ask, concerning our
   third theme, the theme that uncertainty is okay. Now looking back
   over what we have written and what we have discussed, something I
   find very different from other presentations of this material, both pro-
   theism and anti-theism, is that, in general, the message I hear is that
   uncertainty pulls against theism in society. Certainty seems to be associated with religion. Uncertainty seems to be associated with various
   kinds of agnosticism. However, our book is chock full of the idea that
   actually for us individually it's quite to the contrary. The uncertainty
   isn't in direct tension with our religious attitude. In fact, it's part of it
   in some sense. So I wonder if you have any further thoughts about the
   role that plays for you. For you as individuals, how does uncertainty
   relate to faithfulness, as a religious believer and as a follower of Jesus?
AB: That's a good question, isn't it? I sometimes think you need a certain level of confidence in order to be able to live with uncertainty.
   I think if you haven't got sufficient confidence then you're going to
   be fearful, then you dare not take any risks. You need to be absolutely sure of everything, which can lead to a conservative course of
   action or it can even lead to paralysis. Whereas if you've got a good
   confidence, and it may be a confidence based on experience, partly
   your own experience, partly the shared experiences of others, then I
   think you can have the courage to enter the unknown and be an
   explorer. To say, I am going to go and find out about things that I am
   not so sure of and are currently uncharted territory for me, and
   maybe uncharted territory for other people too.
AS: Yes, I agree that that notion of confidence is an important part of
   the territory that one is in when we're thinking about having uncertainty. I think there's a very interesting combination of experiences
   and thoughts in this area, and that the word 'faith' is also part of this
   combination of confidence and uncertainty and so on. An example of
   what I might mean by the word faith is where you have, at some level,
   a confidence that the person you have chosen to follow or the course
   you've chosen to try out is somehow going to work out, it's going to
   be okay. But you do not know that. It's a very interesting combination.
AB: For some of my friends who are in business, the notion that you
   could live without uncertainty is simply foreign to them. They're

living with uncertainty every day, and the successful businessman is the person who can take decisions based on incomplete information, and time and again make good decisions.

HH: I'm curious whether your interaction with these business people illustrates a certain idea that informs science as well. Both of you probably know there's a very famous essay by an English mathematician, William Clifford, concerned with the ethics of belief, where he says that it is one's duty to always proportion your belief to the evidence. He gives his famous example of this ship-maker who wasn't attentive to safety measures, choosing instead to assume that no check was needed, and the ship sinks and it's their first sea voyage and that person is morally culpable. So Clifford claims that the morally responsible person always proportions his belief to the evidence. But it seems to me that what you're saying is that in the case of a good or shrewd business person, often, even within the bounds of being ethical, confidence can exceed the evidence you have. It's not so much an issue of believing in the face of the evidence as much as a commitment, perhaps to the people with whom one is doing business. It's not saying (which would amount to a contradiction) I have more evidence than I actually do. It's saying I'm committed despite the fact that the evidence hasn't forced me to be committed, or something like that.

AB: I think you're right and, indeed, I think that here the commitment is an integral part of the confidence being well-founded. That is to say, the very commitment itself diminishes the uncertainty in the whole situation and transaction.

HH: Yes. As is often the case with personal relations. The knowledge you can gain of a person passively is so much less than you can get actively, by being committed to interacting with them.

AB: Right.

AS: I sense that this part of our discussion is one in which we do need to tread carefully because I don't want to make the mistake of saying or giving the impression, 'I'm just a jolly sort of a fellow who just goes along with what I feel will work out.' Because that isn't how it is either. That would be morally questionable. And it is quite subtle this area. I'd like to add two illustrations from my experience. One is a nice illustration from a piece of creative writing. It's from the film called *Shakespeare in Love*. I have forgotten the author.[1] It's a nice story. It's a nicely put together plot. And in that there's the character of the director of the theatre company who signed up and marshalled together actors and a building, and a financier, to put on a play. And it's all very unclear as to exactly whether it will all come together or not. He has this phrase that he says a few times, something like, 'it'll work out. It always does. It's a mystery but it always works out'. And to me, what we're talking about here is a bit like that. There's a sense that that person in the film can't claim to know that for sure. But he's got enough experience of working in theatre to know how things come together, that the players and actors basically want it to happen. The money will be found. Don't worry, it'll be okay. And it somehow is. Even though he remains somewhat in awe of the process, even though he doesn't quite know how the whole works, it does. That to me captures an aspect of what faith is about in a nice enjoyable way.

But the other thing I'd like to add is an experience that I've had where one begins cautiously. One is nervous and uncertain and one goes through a period of intellectual difficulty as you struggle with questions. Questions such as 'Is what I've been told about how I should receive the Bible—is that the right approach? How do I handle that?' Or questions about the difficulties that there are just in living our life and the pain of animals and so on. You wrestle with these things and it's hard work. But I think my experience has been that nevertheless you do find, or it can happen—it's happened for me—that you get to a calmer place where your faith is no longer in your own ability to answer all these questions. It's more in the One to whom you're looking, the One who is worthy of the faith you are trying to place there. You don't altogether know that One. Not completely, never thoroughly, but you're beginning to have a sense of: they can be relied upon. And you can say, 'I am willing to say that with God things will work out.' That's what the opening phrase of the creed means, I think. 'I believe in God' does not mean 'there's a being that I previously defined whose existence I affirm'; it doesn't mean that. It means: all things considered, after all those anxious things, I think God will see us through; we can place our trust there. Even though we're not altogether sure about all the characteristics and qualities that we are affirming, but nevertheless that's where they are. That's where I come to that place of confidence, you see.

The conversation continues in Chapter 20.

---

[1] The screenplay was written by Tom Stoppard and Marc Norman.

3

Religion, history, and philosophy

The main purpose of this chapter is to introduce religion in broad strokes, and comment on the difference between theological reflection and philosophical analysis. It is a mistake to confuse them. It is also a mistake to think that either is superior to, or can replace, the other.

3.1 Religion

It will be helpful if we use the word *religion* in this book, but since this word has a variety of usages in the contemporary world, we need to define what we mean by it.[1]

There are, broadly speaking, three ways in which the concept 'religion' is commonly understood. This leads to three mutually incompatible definitions.

In one version, one prepares a rubbish bin which represents various shoddy ways of going about life and thought, and one writes the label 'religion' on the side of this rubbish bin, and then one announces that one is opposed to religion. In this first definition, 'religion' is said to be those thoughts and behaviours that people adopt when they choose to be gullible or superstitious or lazy about reason or put themselves in thrall to authority figures.

In the second definition, religion is a collection of ideas and ways of behaving whose main theme and motivation is the attempt to make ourselves acceptable.

In the third definition, religion is that part of our response to life which is to do with discerning the meaning and value of things, and discovering what can properly call upon our allegiance, and in view of this, thinking truthfully and acting well.

---

[1] A glossary of the way that we use some terms in this book is given in Section 3.4 at the end of this chapter.

Depending on what one means by religion, then, one either is resistant to it or welcomes it. But whichever definition one adopts, the various aspects of human life captured by the various definitions of 'religion' will remain, and we need to talk about them somehow. It is simply not useful to change the meaning of words, and it merely generates confusion when different groups use the same word for opposite or very different things. Therefore, we (the authors) want to resist the attempt to redefine 'religion'. We use the word mostly in its positive sense, which is, roughly speaking, the third definition we just alluded to. However, we also acknowledge that much of what goes on under the label of 'religion' is bad.

The mixed bag that religion is in practice has made people want to distance themselves from it. For this reason, it has become common practice for many followers of Jesus of Nazareth to say that what he offers is not religion but the antidote to religion. In this case people are adopting the second definition, in which it is about trying to make oneself acceptable, and what they are affirming is something other than that. They affirm that we respond to our ultimate context appropriately when we do so out of a sense of free acceptance on the basis of love and forgiveness, not on the basis of having made ourselves worthy. Such a response is what in this book we shall call 'good religion'.

In practice, religion is a human phenomenon and consequently not capable of brief definition. The present discussion stands as a way of alerting the reader to this complexity, but also of encouraging the reader to allow for a generous meaning of 'religion', as much as possible.

In the opening chapter, and just now, we have introduced the way of life called 'Christian' by making a connection to Jesus of Nazareth. When we do this, we are speaking of a specific resident of planet Earth who lived, worked, and taught at a certain time and place. However, we are perfectly well aware that the knowledge we have of his life at that time and place—his actions, teaching, and experiences—cannot be completely separated from the experience of the people who wrote about him. Furthermore, our own ability to interpret him correctly would be compromised if we did not have access to the efforts of the community that sprang up in the wake of his contribution. Therefore, we will, in this book, make use of some standard Christian terminology. In particular, we will use the title 'Christ', and in the interests of clarity it will be helpful to say a little about this here.

The title 'Christ' is partly a reference to the history of the Jewish people from ancient times; it places Jesus in that historical sequence and asserts that he fills a role hinted at in the learning process described in the older part of the Bible. This is a role that regards the Jewish community as a humble leader called to serve the rest of humankind. In Jesus the role of servant king was brought to focus and the benefits were universalized. Or so Christians assert by affirming the title 'Christ'. The word itself comes from the Greek *Christos*, and this reminds us that Christianity sprang up in a cultural setting with both Greek and Jewish influences. That same title also refers to another subtle idea. This is that Christian experience blurs the distinction between the man himself and the community of his followers. That is to say, we find ourselves participating in a role that surprises us even as we enact it; somehow we are meant to, and do, in an awkward, imperfect, way, enact in the world a life which is the life of Christ. That is a deliberately paradoxical saying; it is a way of getting at a mode of human existence in which the self is more and more forgotten, but without foolishness or laziness. We also find Christ in some sense living right there in front of our eyes, in the poorest or most afflicted people: hungry, imprisoned, naked to the bruising injustice of the world. Finally, the title signals a meeting point of Creator and creation, a coming together which we will not expound on here because we cannot do justice to it in a few words. Our book is a contribution to a wider effort to show what Christianity means. Our present purpose is best served if we limit ourselves to showing just that small part that we can.

Our world is simultaneously secular and religious. Where once families would say grace before a meal, now those families that eat together might borrow the French *bon appétit* or the American *enjoy*. Few board meetings start with a prayer, or ask what God wills about a decision to be taken, and few universities ask for God's wisdom in their management decisions or curriculum development. Many people, perhaps the majority, in the Western world at least, live their lives with little conscious reference to God. And yet if you open any newspaper, or watch any television news programme, it is not long before you have to have some understanding of religion to make any sense of events. The impact of 'bad religion' is most noticeable in the acts of violence and terror that dominate the attention of the media and their customers. But the effects of religion are also on show in the mosaic of acts of caring and kindness which are the stuff of many non-governmental organizations, and which contribute so richly to making life worthwhile. How can we give an account of the world that adequately takes into account both of these aspects?

One starting point is to observe that humans manifest a curiosity that is sometimes quenched but never exhausted. Humans, like non-human animals but on a vastly higher level of articulation, are curious about the material world, and they also have questions about meaning, purpose, values, truth, reality, and so on. Some of these questions relate to belief in god (written here with a lower case g, for reasons which will be explained later), and sometimes they find a manifestation in religious beliefs and practices.

This kind of curiosity seems to be as old as recorded human activity. There is evidence of burial practices in Africa from nearly 100,000 years ago, and a new expression of curiosity comes from cave paintings from 30–40,000 years BP. An extraordinary expansion of the human prefrontal cortex took place about 200,000 BP, and no one knows why the resulting potential for brain plasticity with its capability for intellectual processing took more than another 150,000 years to manifest itself. Both in Europe and in Africa, and more recently in India, rock art has been found in caves that were not used for living in but for religious ceremonies. The paintings were neither decorations for the living room nor a visual menu for the dining room, but rather expressions of religious resonance, as though they were an early, pre-writing expression of a kind of spiritual yearning.[2]

An appropriate application of the word *religious* is to describe this quest. The object of the quest may be described as god, but for some that may sound too specific, at least for the stage in their quest that they are at, so they may prefer to speak of spiritual reality. Surveys show that many who don't describe themselves as religious are nevertheless content to identify as spiritual.[3] They might believe there is a purpose to the universe and to life which we can try to learn about and enact.

If we follow this approach, then we can expect different lines of enquiry sometimes to converge on the same answer. We can also expect different lines of enquiry to give different insights that illuminate different

    [2] Roger Wagner and Andrew Briggs, *The Penultimate Curiosity: How Science Swims in the Slipstream of Ultimate Questions* (Oxford University Press, 2016).
    [3] E. H. Ecklund, *Science vs. Religion: What Scientists Really Think* (Oxford University Press, 2010).

facets of the truth. The Greek community in Miletus in the sixth century BC developed the idea that there was a first cause behind everything. They called it *arche* (as in words like archaeology—the study of origins). Although there was lively debate about how to describe *arche*, it was a hugely fruitful concept because in place of explanations of natural phenomena as acts of gods it gave space for investigation into causal mechanisms.

Different religions can be seen, in part, as different traditions of inquiry into what god is like. Insofar as they are trying to describe the same reality, it is to be expected that they would come up with similar, or at least complementary, insights. Thus, we might expect some common sense of right and wrong, and of the roles of justice and forgiveness. We can expect that all religions would recognize a causal order in the way the world works, and at the same time a recognition that bad things happen to good people. We might also expect a shared sense of the greatness of god, and an awareness that it is appropriate to worship god and to pray to god.

We have written of god with a small g so far, because we have written about human curiosity and attempts to seek. The widespread human experience of seeking for god does not necessarily mean that we understand correctly the nature of what is being sought, just as the experience of thirst does not mean that one already knows what water-based liquid is, but in each case the appetite is consistent with there being a valid object of the desire, and the appetite is suggestive; it invites us to explore. Let us suppose now that there is indeed a valid object of this human desire. This is a large step, one which some readers may have already made and be eager to pursue further, but we would like to keep the discussion open to everyone. So for other readers we are now simply inviting you to look at the situation from a certain point of view, with the aim of seeing what that viewpoint is like, or what would follow. What would it be like if the reality underpinning the physical world and somewhat expressed in it, or that which is the reason why there is a physical world, is rich in attributes and has taken some sort of initiative to engage with humans? If so, then we must make a distinction between god as a word to describe a concept which one can consider and God as a name for a being to Whom one can relate. We are using the capital G now as a way to raise awareness that the situation changes when it involves a meeting with our whole self, as opposed to merely satisfying curiosity.

In this view religion changes from being a human initiative to a divine initiative. The emphasis shifts from humanity's quest for god to God's engagement with humankind. It shifts from being an academic exercise (although there is plenty of room for that) to being an engagement. The engagement is told through a narrative. That is why it is useful that God has a name. When Moses was charged with rescuing the Hebrews from a tyrant, he wanted a name for the one who commissioned him. The name had the Hebrew consonants YHWH, which out of deference was pronounced Adonai, the Hebrew for Lord,[4] and was in due course annotated with those vowels. The word itself is an expression of the verb *to be*, and can be translated 'I am' or 'I am who I am' or 'I will be who I will be.' It did not function as a label in any brief way, but rather as an invitation to learn. The description of God became inseparable from the history of people who shared their experiences of God. God is the God of Abraham, Isaac, and Jacob. In the experience of the Hebrew community, the One Who is being described is One Who self-identifies through a covenant, which is to say, through relationships of trust, through people and their relationship with that One.

Throughout this book, we shall occasionally capitalize a metonym for God, as writers in previous ages often did. This is not out of some quaint sentimentality, but as a reminder of the first theme which we introduced in Chapter 1, that God is a being to be known, not a hypothesis to be tested. We want to be liberated from devising some human concept of god, and then trying to discover whether god thus defined exists. This book is all about seeking what or who is there. If we write 'what', it might be taken to mean that we think that God is a thing. But by writing 'Who' we would not want to prematurely anthropomorphize the object of our seeking. We use the pronoun 'He' for brevity; we acknowledge that gendered language is not altogether satisfactory, but we find this to be better than impersonal language; we have to work within the limitations of contemporary English. Rather as the Hebrew scriptures used YHWH, so we shall sometimes use capitalization to give identity to Him Whom we seek. Even that language is barely adequate, because we are writing about the God Who is there, and we want to be open to finding that He took the initiative in reaching out to us.

---

[4] אֲדֹנָי literally means 'My Lords'.

3.2 Philosophy getting muddled about religion

We have been careful in our use of the word god and God. We have not described the religious aspect of life as a journey in which we discover whether there really is someone called God, but as a journey in which we discover what there really is. In such a discovery we learn what names are appropriate. It is not for us puny humans to pronounce judgement on the very existence of that which causes our own exist-ence. That would be a hopelessly muddle-headed thing to do. Instead, we make it our aim to learn what is the character of that which makes our life possible and can enrich the lives of others around us. Unfortunately, Western philosophy has often failed to appreciate this distinction. In much philosophy, the word 'god' has been employed to signify some sort of entity that might not exist, but whose existence we can try to demonstrate on the basis of some axioms and reasoning. This muddle-headed attitude may be a symptom of hubris, or simply of genuine confusion, like someone trying to use logic to prove that logic is itself valid. It seems to have its origins in a misinterpretation of High Medieval European religious thinkers.

In the Middle Ages, around the twelfth and thirteenth centuries, classical Greek thinking reached Europe via Arabic translations. It is hard for us to appreciate the intellectual shock waves that hit thinking people of the time—and there were plenty of them. There were two significant components to their response. First, they sought to define the object of their inquiry—god—in terms of the rediscovered Greek thinking. Such a definition might include properties such as omnisci-ence, omnipotence, and infinity. One thinker, Anselm, wrote of god as 'whatever it is better to be than not to be'. He went on to ponder the fact that god is said to be *that, greater than which cannot be conceived*. Second, they sometimes sought to prove the existence of god from basic axioms, rather as Euclid proved that the angles of a triangle add up to 180°, from axioms about the sum of angles on a straight line and angles associated with parallel lines. Thus, Anselm argued from carefully chosen axioms or premises to the conclusion that god must exist. His first premise was that 'god is that "X", greater than which cannot be conceived'. His second premise was that 'that which exists is greater than that which does not exist'. It follows, says Anselm, that god exists, since a non-existent god would not be *that greater than which cannot be con-ceived*. This way of arguing immediately felt unconvincing and met with resistance, and indeed Anselm himself wrote mainly of his experience of struggling to know God, rather than saying he had solved the mystery. The logical fallacy of the argument attributed to Anselm has been elaborated by subsequent thinkers. From our perspective, we would say the problem lies in asking an ill-posed question and thinking of god as a thing whose existence can be debated. But it may also be because we misunderstand the culture in which such arguments were being discussed.

The most famous, because the most influential, thinker of the time was Thomas Aquinas. To the modern reader, it might seem that he was trying to prove things about god that his readers did not believe, but that would be a mistake. Rather he and his contemporaries were seeking to show that their religious faith could stand up to the most rigorous intellectual scrutiny. That is to say, Aquinas did not set out from a position of denying god and then using arguments to show that one must accept god after all. Rather, he set out various ways of getting at what it is we mean by god, and showing that they were reasonable. Aquinas's ideas are sometimes presented in university philosophy courses as 'proofs for the existence of god'. That is to misunderstand what Aquinas was trying to do, and to invite students to engage upon a thoroughly misconceived intellectual exercise, one which starts out by giving a highly artificial meaning to the word 'god', using definitions that one can invent and inspect passively. Instead of the wonderful Hebrew idea, the name which is an invitation, connoting *that which is* and *that which can be discovered*, students are invited to place themselves in the role of judge and executioner of a passive prisoner, already half-dead, that they have caught in the net of their definitions.

A more recent instance of a similar phenomenon, although already nearly two centuries old, is associated with reactions to the work of William Paley. Paley wrote in great detail about the way that animals are superbly optimized for their ecological niche. Fish are streamlined for swimming, birds are aerodynamic for flying, and so on. He gave many examples. Darwin said that he learned much as an undergraduate from studying Paley's *Evidences*. Paley also used an argument by analogy with a walker on a moor coming across a watch on the ground, where the intricate mechanism of the watch would be evidence of the craftsmanship of the watchmaker. Just so, he explained, the details of the living world provide evidence of the creativity of god. Paley did not have the advantage that we have of an understanding of the process of evolution through natural selection. But he did live in an intellectual world in which for over a century people had been pointing to the marvels of astronomy as evidence of god's creativity. Robert Boyle, in the seventeenth century, had even used a famous clock in Strasbourg as an illustration. Paley extended what had previously been argued from the mathematical world of planetary motion and physics to the (at the time) very unmathematical world of natural history and biology. If we now find his arguments unconvincing, it may be because we misunderstand the job that he was doing and the intellectual culture in which he was doing it.

Those familiar with the Hebrew scriptures will recognize a similar pattern there. There are many instances where the Hebrew Bible refers to observations about the material world, most notably in the Psalms and the Wisdom Literature, but also in the Prophets and the opening chapters of Genesis. But never, we think, do any of those writings use observations about the creation to argue for the existence of a god whom the reader might otherwise think unreal. Rather, they use what we might now call scientific observation to illustrate aspects of God's character that the reader might otherwise struggle to understand.

3.3 The divorce settlement?

Stephen Jay Gould (1941–2002) was a distinguished palaeontologist, evolutionary biologist, and widely read science writer and historian of science. Working mostly at Harvard University and the American Museum of Natural History in New York, he was thoughtful about the tension in American public life surrounding the teaching of evolution in schools. He proposed a view which tried to eliminate any possibility of conflict between religion and science by introducing the concept of non-overlapping magisteria—which he called NOMA. He explained the concept thus:

> the magisterium of science covers the empirical realm: what the universe is made of (fact) and why does it work this way (theory). The magisterium of religion extends over questions of ultimate meaning and moral value. These two magisteria do not overlap, nor do they encompass all enquiry (consider, for example, the magisterium of art and the meaning of beauty).

He later recapitulates:

> NOMA is a simple, humane, rational and altogether conventional argument for mutual respect, based on non-overlapping subject matter, between two components of wisdom in a full human life: our drive to understand the factual character of nature (the magisterium of science), and our need to define meaning in our lives and a moral basis for our actions (the magisterium of religion).[5]

Professor Gould correctly identified the important distinction between knowledge and wisdom, facts and meaning, process and value. However, the picture he describes fails in two important respects.[6] It fails first because it sets forth religion as one of several legitimate spheres of enquiry. That may be adequate for some part of religion, including the search for god with a small g, rather as theology can serve as a useful academic category, but it is far too constraining as a bound for religion considered as our response to the whole of God's engagement with humankind. Secondly, it fails even more disastrously by excluding facts from the purview of religion. If God engages with humankind, then that engagement is made up of a series of facts arranged in a narrative.

In the eighteenth-century period called the Enlightenment,[7] the experiment of a religion devoid of revelation was tried. On a large scale, the experiment consisted of having a religion based solely on what could be deduced from nature, starting with physics and biology in the seventeenth century and ending with geology and biology in the nineteenth century. The approach was captured in an ode by the journalist and editor Joseph Addison.

> The spacious firmament on high,
> With all the blue ethereal sky,
> And spangled heav'ns, a shining frame,
> Their great original proclaim:
> Th' unwearied Sun, from day to day,

---

[5] Stephen Jay Gould, *Rocks of Ages* (Ballantine, 1999), 6, 175.

[6] G. A. D. Briggs, 'Stephen Jay Gould, *Rocks of Ages: Science and Religion in the Fullness of Life*', *Science & Christian Belief* **12** (2000), 177–80.

[7] The intellectual liberties won in this period are rightly celebrated, but the period is also associated with a loss or diminishment of other goods. These are concerned with allowing a sense of community and common purpose to inform our conception of human identity, refusing a too starkly individualistic model. Also, we need to reaffirm our freedom to explore and express what does not contradict reason but is not completely captured by reason.

> Does his Creator's pow'r display,
> And publishes to every land
> The work of an Almighty Hand.

This is an advance on NOMA, because it does at least allow for observations, and hence facts. But it denied any direct communication from God, beyond the initial creation. This was articulated in the final stanza of Addison's ode.

> What though, in solemn silence, all
> Move round the dark terrestrial ball?
> What tho' nor real voice nor sound
> Amid their radiant orbs be found?
> In Reason's ear they all rejoice,
> And utter forth a glorious voice,
> For ever singing, as they shine,
> 'The Hand that made us is Divine.'

Addison's ode was brilliantly set to music by Benjamin Britten in his *Noye's Fludde*. It is based on the first half of Psalm 19. The second half of Psalm 19 uses five different synonyms for written or spoken communication to articulate how God communicates now.

The intellectual movement of which Addison was a spokesman became known as deism, meaning belief in god as creator but without subsequent revelation, as distinct from theism, meaning belief in God as both creator and as One Who sustains and engages with the universe in personally significant ways. The demise of deism marked the end of an experiment in religion that failed, or, if you prefer, gave a negative result. The great scholar of science and religion Ian Barbour, who in many ways was the twentieth-century father of the intellectual study of that topic, summed up the result thus,

> The waning of Deism can be attributed primarily to its own inherent weaknesses. The Cosmic Designer, who started the world machine and left it to run on its own, seemed impersonal and remote—not a God who cares for individuals and is actively related to man, or a Being to whom prayer would be appropriate. It is not surprising that such a do-nothing God, irrelevant to daily life, became a hypothesis for the origin of the world or a verbal formula which before long could be dispensed with completely.[8]

[8] I. G. Barbour, *Issues in Science and Religion* (SCM, 1966), 61–2.

Although Earth's rotations take place in silence, this does not mean that God does not speak. The observations of astronomy can help us to enlarge our imagination and our sense of God's beauty, mystery, and greatness, but these intuitions need some more specific information to work on. That more specific information has come into the possession of the human race through an interaction involving a trusting relationship, on both sides, rather than merely viewing the results of God's creativity. In the narrative of this relationship of trust, or covenant, built gradually over a long period, there is a form of communication which has been given the theological term *revelation*.

Revelation is itself an idea which has some nuance, but essentially it is the claim that there is an aspect to our relationship to God which is like the relationship we have to one another, in which in order to get to know one another we must open up a little and let ourselves be known, and we also must take an interest in coming to know the other. This is done mostly by talking and by spending time with one another, doing things together. Something like that has happened in the journey of the people of ancient Israel, culminating in the Rabbi who interpreted for them, and for us, what that revelation was really all about.

Such comparisons or analogies between human relations and the relationship to God never succeed completely, but they are an important part of the complete picture. This is why they should not be ignored, as the deist movement tried to do. Equally, though, one should not propose the simple equation 'revelation = the Bible' because such an equation is not sufficiently receptive to the types of wisdom which the Bible itself promotes. Rather, revelation took place through a process. The Bible is the record, from the human end, of what the process felt like to the people involved, and what they gradually learned.

Andrew Briggs recounts:

My grandfather, George Wallace Briggs (1875–1959), was a Canon of Worcester Cathedral. He wrote a widely used hymn to celebrate God's speaking through His prophets, through Christ Jesus, and through his Spirit. An 'unchanging word' is thus revealed. Revelation begins with a sense of foundational justice, that despite the injustices of the world there is a higher court where the truth will out. Then we see God's character revealed more intimately in a human life. Finally, we receive the gift of being able to appreciate these things and live accordingly. The thought that God's word is unchanging is not restrictive, because it is like the way the laws of nature are unchanging. It is because the natural world is dependable that we have liberty to do something creative in it. The unchanging Word that we receive is the announcement that we come to God on the basis of trust and free forgiveness; that God is all about welcoming the prodigal and joining with the peace-makers and discovering the value of the people whom others despise.

In considering religion as a human quest for god, there is plenty of scope for complementary insights; that is why it is fruitful to seek wisdom in all the world's religions. But the transition in thinking to God's engagement with the world and its inhabitants makes the data more specific, and therefore open to differences. The Cambridge professor David Ford uses the metaphor that followers of different religions can invite one another into their traditions rather like guests into their homes. The hosts do everything they can to make their guests welcome, but they do not move walls or change furnishings to suit the preferences of their guests. The guests accept hospitality on those terms. They are grateful for the welcome that they enjoy and for the hospitality that their hosts have chosen to share with them, and they expect to have only a limited say in what they will receive.

In my experience, when I have participated in meetings with people of other religions, especially the Abrahamic faiths, I have sometimes started by expressing appreciation for what we can agree about in our common search for truth. But I believe (and I generally say so) that we show respect for one another by being willing to articulate what is distinctive in our own faith. Time and again I have found that if this is done courteously, then it is received appreciatively. On one such occasion, when the majority present were of Islamic faith, I described how the account of the logos becoming physical opens up for me a whole new understanding of God's engagement in the world, by being literally incorporated in the highest form which we humans are capable of relating to. The response to this was warm, acknowledging the benefit of being able to see in sharp focus what can otherwise appear blurred or vague.

3.4 Glossary

We present here a glossary of some terms which we use in specific ways. The commonly understood meaning is not to be taken as the universally understood meaning, nor as the meaning that would be given by scholars; rather, it is intended to contrast the way many people might use the term with our use in this book. The meaning used by us is not a full operational definition, but rather a short approximation of what we explain in more detail throughout the book.

Word or phrase	Commonly understood meaning	Meaning used by us
God	A being whose existence can be argued about (god)	The fullest meaning of the verb *to be*; One Who can be known
religion	A set of practices in relation to beliefs about God	A way of living in relation to God, usually within an established tradition
theology	A set of beliefs or speculations about God	A way of discovering knowledge about God, involving grappling with all evidence and experience
science	A body of established knowledge, as in 'science tells us that …'	A way of discovering knowledge about the material world, usually involving a combination of theory and experiment or observation
creation	The coming into being of the world through specific divine interventions, sometimes considered to be relatively recent	The bringing into being of the universe and everything in it, through processes which can be reliably studied scientifically
nature	An autonomous source of life and being, sometimes referred to as Mother Nature	The material world, practically synonymous with the creation
the Bible	An infallible record of the word of God	The definitive record, within the Christian tradition, of human experience of God
soul	A distinct component of life, which can survive after death	A living organism, usually human (as in 'there were 100 souls on board'), with particular reference to what defines their character
faith	Forming beliefs without evidence	Willingness to respond to suggestive evidence
Christianity	A historical, organized religion	Learning from Jesus *Christos* and the community of his followers, and coming to know him in response.

4

How is science to be carried forward, and its conclusions reported?

4.1 Three reasoned arguments and two ways of reacting

When one of the authors (AS) was a teenager, his elder brother showed him the proof, due to Euclid, that there are an infinite number of prime numbers.

The proof runs as follows.

We suppose first that there is a finite number of prime numbers, in order to arrive at a contradiction. On the supposition that there are only finitely many prime numbers, there must be a largest prime number, N. Now consider the number x that is given by one less than the product of all prime numbers up to and including N. That is,

$$x = (2 \times 3 \times 5 \times 7 \times 11 \times 13 \times 17 \dots \times N) - 1.$$

x is clearly larger than N. If x is itself a prime number, then our claim that N was the largest prime number is untrue, which is a contradiction. Therefore, x is not prime, which implies that x must have prime factors. But we have defined x in such a way that it is not divisible by any of the prime numbers up to and including N. Therefore, any prime factors of x must themselves be larger than N. This once again contradicts the assumption that N is the largest prime, and there is no way to avoid this conclusion. It follows that our original assumption, that there is a highest prime number, must be wrong. Q.E.D.[1]

The reason why we began the chapter with this proof will emerge as we go on.

Let's consider next a piece of basic physics.

---

[1] Q.E.D. is an initialism of the Latin phrase *quod erat demonstrandum* meaning 'what was to be demonstrated'.

A model of physics which succeeds to very good approximation, when dealing with planetary motion, is the one now called Newtonian. In this model, we say that every massive body attracts every other massive body, with a force given by Newton's formula of universal gravitation:

$$f = \frac{(G\, M_1 M_2)}{r^2}.$$

In this formula, f is the force, G is a measured constant called the universal gravitational constant, M_1 and M_2 are the masses of the bodies involved, and r is the distance between them.

Also, in Newtonian physics, a body in motion will continue in motion at a fixed speed and direction unless forces act on it. If a force does act, then the momentum or impetus (the product of mass and velocity) of the body will change at a rate equal to the size and direction of the force.

By bringing these ideas together, we find that the velocity of Earth, on its journey through space, is influenced by the force of gravitational attraction to the Sun, in the following way. The *change* in Earth's velocity, in any given small time t, is

$$\frac{t\, G\, M_S}{r^2}$$

where M_S is the mass of the Sun, and r is the distance from the centre of the Earth to the centre of the Sun. In the case of Earth and the Sun, the change in Earth's velocity, in any given small time, is mostly in a direction at right angles to the direction of travel, which results in the Earth following an almost circular path.

This equation now forms a standard part of first-year undergraduate courses in physics, where students learn to deduce, by mathematical analysis, the precise behaviour of the velocity, and hence also the trajectory followed by the Earth. If one assumes the Sun is fixed, then one finds that Earth moves around a nearly circular ellipse, and one can relate the size of this ellipse to the speed of travel around it. Upon measuring the distance from Earth to the Sun by astronomical observations, one can then deduce the time it must take Earth to go once around its orbit. One finds that the time thus deduced or predicted is one year. But that is also what is observed empirically. Therefore, there is agreement between observation and theory, and this is evidence that the whole model holds good.

A major objection to this Newtonian model, one which Newton was himself aware of, is that it says nothing at all of how the force of gravitation comes about. It merely asserts that the force is there, and that it is regular, being always given by the same combination of masses and distance and a universal constant. The scientific community has, in the twentieth century, discovered a more accurate model of gravity, the one given by general relativity, which gives a more detailed model of how gravity comes about, but in the end even this more sophisticated theory has to rest content with saying merely 'this is so' about its basic concepts and equations.[2]

Again, the reader must bear with us before it becomes clear where we are heading with these examples.

Next, we turn to an example in biology.

A group of bird species that are found in the Galápagos Islands have come to be known as 'Darwin's finches'. The term was adopted by Percy Lowe and popularized by David Lack in his book of that name. The species of birds are in many respects similar to one another, but have very differently shaped beaks. Darwin began the organized study of these birds, writing in his 1845 second edition of *The Voyage of the Beagle* (now entitled *Journal of Researches*):

> Seeing this gradation and diversity of structure in one small, intimately related group of birds, one might really fancy that from an original paucity of birds in this archipelago, one species had been taken and modified for different ends.

He developed this thought in *On the Origin of Species*, where he notes various pertinent observations. These include that the island species are like those on the nearest mainland, without being actually the same species. Also, that the geography and climate of the Galápagos Islands differ from that of the nearest mainland (several hundred miles away), while being similar to other islands such as the Cape Verde archipelagos, but the birds on those other islands are quite unlike the Galápagos species. This suggests that the birds did not arise directly on the islands (for then we would expect similar islands to have similar birds). More likely, ancestors of the Galápagos birds arrived there from the mainland

---

[2] However, when one studies general relativity, one does begin to develop a hint of the feeling of not just 'this is so' but also 'this is wonderfully elegant; could it be that it must be so?'

(where there are similar species), and then the new species (the ones now present on the Galápagos Islands) diversified from those ancestors.

That this is what happened is no longer in doubt. It has been further studied, and many more details have been looked into. Some of the genetic basis of the beak structure has also been established.[3]

Now we come to the point we wish to make in this chapter.

We have just presented three examples of reasoned argument: Euclid's proof of the infinity of primes; the mathematical/physical model of the motion of astronomical bodies; and the process of diversification in living species. In no case did we make any explicit mention of God. It was not necessary to introduce a step in the middle of Euclid's proof, in which one asserted that God would guarantee or arrange some mathematical result which might otherwise be doubted. In the Newtonian model of gravity, it was not necessary to say that God acted to keep Earth in motion, or to make gravitation happen. It was sufficient to say that Earth did stay in motion and gravity did act. In the sequence of diversi-fication of species, it was not necessary to say that God created new species. It was enough to say that new species came about by accumu-lated change from old species.

There are two ways of reacting to this situation.

One way is to thank God for mathematical logic, for harmonious astronomy, and for diversifying and struggling life. The other way is to be thankless towards God, either because one is atheist or because one is theist but does not recognize that the marvellous soundness of the natural world is something we can thank God for.

In the first way, one recognizes that what God offers and guaran-tees, in the context of mathematical reasoning, is the role and good-ness of logic itself. To embark on mathematical reasoning is not to depart from God or ignore Him, but to embark on an enjoyment and appreciation of the very thing God is furnishing. Coming now to the astronomical example, there is more contingency: one feels that things could have been otherwise (perhaps a different number of physical dimensions, for example, or a different sort of time), but as it happens they are like they are. So there is more of a sense that it all came from some sort of creative act. But, like mathematical proof, astronomical motions are highly regular. In discovering this regularity and setting it

[3] A. Abzhanov, M. Protas, B. R. Grant, P. R. Grant, and C. J. Tabin, 'Bmp4 and Morphological Variation of Beaks in Darwin's Finches', Science **305**/5689 (2004), 1462–5.

forth, one is not insulting God by failing to mention Him at some point *within* the system of forces. Rather, to be faithful to God is to see and describe the system of forces and motions in their own terms, because that is the very system that God has furnished and guaranteed. The gratitude comes at the end, when we sit back and contemplate this great system, and experience the sense of marvel and not a little awe.

The third example, living species, is the one which causes the strongest feeling. There is a much greater contingency, and a much greater degree of sophisticated pattern, and more chance. Faithfulness to God certainly must take the form of seeing the natural world as truthfully as we can. Are we allowed to declare that God made the elephant and the orangutan and the humble bumble bee? Yes, but not in such a way as to intrude illogically into the middle of a scientific study of the sequence of processes that have, in fact, been involved. Just as with the other cases (mathematical logic and models of physical motion), one must respect and celebrate the object of study in terms appropriate to its own nature. This is the way to respect and celebrate God Who gives all things the chance to be what they are.

It is appropriate to acknowledge also that the natural world is not exempt from morally objectionable pain and strife. This is because the world is not divine and is in some respects radically free. The world has some aspects that thrill us, and others that dismay us and summon up our commitment to act in the world, to change it. We should therefore be cautious of sermons or song lyrics which present the natural world as completely beautiful and harmonious, i.e. as other than it is. Our songwriting model should be, arguably, the Psalms, which display honesty about natural processes, and a strong note of lament alongside joy and celebration.

We stated earlier that there were two ways of reacting to scientific study, but so far we have only discussed one. The second way is the way of atheism. There is perhaps a certain irony involved here, because the more we do science and become able to appreciate the functioning of the world, the more we become aware of its self-consistent integrity. Hence, the more we develop our scientific understanding, the more we may begin to see the physical world as autonomous, with ourselves part of it and wholly unrelated to God. Does this make scientific understanding bad for us? Surely not! Could it be good, then, that we come to experience such feelings of autonomy? Even the autonomy of declaring, in complete honesty, 'I think the measurable, tangible, natural world is all there is. Here I stand; I can do no other'?

This brings us to an important principle that is at work in all relationships of persons, and it is especially important in our understanding of our position before God. The issue is, are we loving God only for what He can do for us? Are we turning to Him only because we were brought up to think that the door to atheism does not stand open? Are we frightened of science because it may open that door, or open it more fully?

But if loving God does not stand for a way of being human that is better than atheism, then our faith is worthless already, science or no.

The book of Job in the Bible offers a profound reflection on what it means to acknowledge God. In the prologue to the book, the righteous man Job is placed under an accusation by a suspicious prosecutor: the accusation of loving God only for what he can get out of it. The book then presents a lengthy discourse as Job's friends and acquaintances dispute with him their interpretations of his situation as he is allowed to be afflicted with a sequence of terrible painful experiences. Job refuses to accept his so-called friends' claims that these experiences are his own fault, and he also refuses to reject God, but he does feel that God owes him an explanation. Job is finally addressed by God in a scene which is sometimes interpreted as an almighty put-down, but we find, on the contrary, a touching intimacy as God lifts Job's eyes to the wonders of the created world, and Job is vindicated by the very fact that God wants to talk with him and enlarge his vision.[4]

Jesus told a parable that also addresses our mixed reactions when it comes to our standing before God. The parable addresses the issue of autonomy and invites us to understand God's response to our desire for autonomy in a stunning way. In the parable (which has become known as the parable of the prodigal son), one of a pair of sons leaves home, taking his inheritance with him and breaking off ties with his father. The young man's life descends into ruin. When he comes to his senses, he reasons that he would be better off even as a member of staff in his father's household, if only the old man will take him in. But what happens is that the father eagerly searches for his son's return and runs to meet him in full and free celebration. The return of the prodigal son is both a deeply human and also a profoundly theological story. It has

[4] T. C. B. McLeish, *Faith and Wisdom in Science* (Oxford University Press, 2014).

been imagined by many artists down the centuries. Two notable examples are works by Rembrandt van Rijn and by Roger Wagner.[5]

God does give us our autonomy. He gives us it radically, right in the core of our identity, if that is what we think we want. Of course, we are not autonomous really, and would not last a picosecond without Him, but like a good parent God knows that we need to come to love Him for Who He is, not for what He can do, and He will allow any avenue that can, in justice, accomplish that. So God's role as guarantor of the physical world is kept carefully neutral, utterly independent of our heart's attitude, and even capable of being seen atheistically when human witness to God is ignored or rejected. Thus, God gives us freedom to turn either towards or away from Him. In this way, and perhaps only in this way, our motives in turning to Him can be pure. In pre-modern eras when it was widely taken for granted that the world comes from God, people did not for that reason love God for Who He is. They had to learn to do that, just like us. In post-modern culture, it is widely taken for granted that there is no God, but the degree of blundering here is possibly no greater than in previous eras. It is merely of a different kind.

This discussion of the natural world's self-consistent autonomy might make the reader feel we are admitting that scientific study leads one towards deism, in which one comes to think that the creator of the world does not care about the world, nor act in it, which amounts to atheism once again. However, our book will, we hope, make it abundantly clear that that is not what we think. It is simply that in this chapter we have, so far, only mentioned those parts of science which deal with inanimate objects and processes. (Living things are not inanimate, of course, but the process of diversification and natural selection is.) It is only when we come to consider personal relations that personal categories come into play. We think that it is perfectly correct to say of certain developments in the course of human life, 'For flesh and blood has not revealed this to you, but my Father in heaven'. This is Jesus' reply to Peter,[6] just after Peter first realizes that Jesus is not just a remarkable teacher and healer, but a man playing a uniquely important role in Jewish history and consequently world history. The insight that

---

[5] The two images can be found, respectively, at https://en.wikipedia.org/wiki/The_Return_of_the_Prodigal_Son_(Rembrandt)#/media/File:Rembrandt_Harmensz_van_Rijn_-_Return_of_the_Prodigal_Son_-_Google_Art_Project.jpg; and http://www.rogerwagner.co.uk/work/item/169/the-return-study-2012

[6] Matthew 16:17.

Peter arrived at is a physical development in the sense that it involves synaptic connections in Peter's brain, and furthermore Peter came to his insight largely through witnessing events and pondering them and opening himself to their possible meaning over an extended period of time. In other words, this synaptic connection came about largely through natural processes. But Jesus is declaring that, in this instance, what has taken place for Peter amounts to something not fully contained in natural process ('flesh and blood'). The absolute Father has met with Peter's open thoughtfulness and responded as one person to another.

It is in this kind of encounter that God makes His presence known to us. It is an exchange of perfect respect, a realized connection as wise as what wisdom we have got, and as deep as we have so far become. One could offer many examples. The opportunity for such a step is generally brought about on our side by paying lengthy attention to what is going on around one, including in wider human experience if one has access to records and accumulated wisdom, and by bringing along a small further contribution, a grain no bigger than a mustard-seed, of willingness. There is, arguably, a less complete but somewhat similar encounter going on in all our steps of learning of more everyday matters. Every correct insight that anyone ever arrived at has been formed by virtue of a meeting with God at some level. Those that we allow to engage us more deeply are the ones that make more space for God to be Who God is.

4.2 Letting science be science: an illustration

The sense of ease and unease has oscillated in the relationship between science and religious commitment. Sometimes the relationship is comfortable, sometimes it is uneasy.

A case study is found in two leading scientists of the nineteenth century.

John Herschel and Charles Darwin were among the most able of the scientists of their time. Herschel, the older man, was an important influence on Darwin. His *A Preliminary Discourse on the Study of Natural Philosophy* (1831) set out clearly and in detail the scientific and inductive approach, and this treatise was highly valued by Darwin throughout his life.[7]

---

[7] Roger Wagner and Andrew Briggs, *The Penultimate Curiosity: How Science Swims in the Slipstream of Ultimate Questions* (Oxford University Press, 2016).

Herschel's scientific work was mostly in astronomy, but he also carried out botanical investigations, and he noted that the question of the development of life on Earth, especially the emergence of new species, was an area ripe for exploration, and ought to be investigated.

Darwin greatly respected the great 'Sir John' and sent him a copy of *On the Origin of Species* when it was published, hoping that it would satisfy his high standards of scientific enquiry. Herschel's initial reaction seems to have been lukewarm, since he was not initially convinced that there could be such a large role for happenstance (he called it the law of 'higgledy piggledy'). In this reaction Herschel had in mind the evident structure both within all living things and in the myriad interrelationships throughout the ecosystem. However, he remained adamant that scientific methods should be used to resolve such questions.

The trajectory of Herschel's religious convictions was from a sceptical position as a young man towards a recognition of God that grew and remained warm in his later years. Conversely, Darwin moved from a form of youthful religious commitment towards greater scepticism in later life, though without arriving at a settled position. Both men were distinguished scientists who were equally well aware of the need to allow scientific study to work within its own domain by its own proper methods.

When Darwin wrote *On the Origin of Species* he was especially sensitive to the need to present a properly argued scientific discourse that would be recognized and treated as such. This is why he was careful not to bring in religious language. Very likely he knew of, and drew a lesson from, a well-known mistake made by Newton, for example, who had affirmed that God acts specially to stabilize the Solar System, as if something other than natural process was involved in the motions of the planets. Darwin could not be sure of the complete correctness of his argument, but he bravely 'held the line' and spurned the 'God of the gaps' approach. In this he was both theologically and scientifically correct.

Herschel could respect all this, but he was inclined to add a note of overall recognition of God, as the provider both of the natural world and of our capacity and opportunity to understand it. Darwin preferred that such recognitions be placed, by those who wanted to offer them, at one step removed from the scientific work itself. He recognized that such statements are, in the end, statements of faith, and not quite the same thing as scientific analyses or deductions. It is not (*nota bene*) that statements of faith are unmotivated; far from it. It is simply that they are in part motivated by assessments of moral value and personal commitment, not just impersonal data.

4.3 Good practice

This question of how to present scientific work in the context of religious commitment is a delicate one. It is like the question of when to applaud at the end of a live performance of orchestral music. If we rush in too quickly with our applause, then we intrude on the moment of silence which is itself one of the most exquisite moments of the music, and we fail to allow the music to be its own whole. If we delay too long, on the other hand, then we are failing in human responsiveness and fellow-feeling.

To take another example: when an artist has produced a great work of visual art, then she wants us to look at the picture. If the picture is unveiled at an exhibition, the artist does not want to be invited up on stage and quickly become the centre of attention, while the picture is ignored. To engage with her work is to come alongside the artist and try to see what she sees; otherwise, we isolate her and render her unable to speak in her chosen language, no matter how well meaning our toasts and expressions of thanks.

So, in scientific work, our first response should be to allow the work to speak for itself in its own terms. We should not insist on a faithful statement of praise to God in the introduction or conclusion, and it is, we think, largely correct that God be not included in the acknowledgements section either, in scientific papers. This is in order to respect the diversity that exists in the scientific community, and to avoid pomposity, and to avoid any suggestion that such acknowledgements have any bearing on the quality of the scientific work that has been done.

On other hand, in the case of a larger piece of work that has more of a sense of personal property, such as a textbook or a doctoral dissertation, then, we think, an expression of thanks to God, done with due recognition of the multicultural context, and with a light touch, should be allowed. It might be located, for example, in the same place where one may choose to thank one's friends or family for their support.

5

What does it mean to be me?

It is a recurring theme of science fiction to explore what it would be like for a computer or a robot to have conscious thought like our own. 'HAL', the computer which is intended to keep the humans safe on the spaceship in the film *2001: A Space Odyssey*, decides to turn against them. In the novels of Iain M. Banks' *Culture* series, there are huge spaceships governed by large machine-built 'minds' whose cognitive capacities far exceed that of any human. Poets and philosophers have also wondered about issues such as change and identity. What does it mean to say that a single continuous person is involved in the experiences and changes undergone by the human body as it grows?

Some of the things we do are instinctive—pulling a hand away from a hot object, craving food and water. Some are personal preferences of no great significance—our choice of drink at the start of the day, our favourite shirt. Although someone who knows us may consider things like that as fondly remembered details, we would not want to be defined that way. Other things matter more to us, such as whether we are considered trustworthy, or a good parent or friend or a capable colleague. Those are closer to the heart of the question 'What does it mean to be me?'

Philosophy has often grappled with this issue, and sometimes works of philosophy can be puzzling or unsettling for people with religious commitments. Well before we had met one another, all three of us (the authors) came across a widely read work of twentieth-century philosophy, *Language, Truth and Logic* by A. J. Ayer, published in 1936.

One of us (AS) came across the book when a nephew was reading it, and noted, from a brief perusal, that Ayer seeks to show, among other things, that language involving the word 'God' has strictly no meaning, on the grounds that it does not and cannot have objective empirical content. So if the argument is right, then Ayer and his philosophical colleagues have at last solved one of the great problems of human thought. What a tremendous achievement that would be! But perhaps things are not quite that simple, because of empirical evidence such as the resurrection accounts, and more generally the evidence of witnesses to God of one kind or another. Also, Ayer's argument leads to the conclusion that there is no objective content to moral statements either, no notion of right or wrong. So if one has good reasons to think that, in fact, there is such a thing as right and wrong, then this may make one suspect that there is some fault in Ayers' argument.

Another of us (HH) came across this same work as part of a standard training in philosophy, and it did not trouble him much because it was quickly pointed out, within the community of professional philosophers, that the argument of the book does indeed fail, and in fact it is not hard to show this.

Finally, the third of us (AB) once found himself seated next to A. J. Ayer at a college dinner. Seeking to turn the conversation to an agreeable topic, AB (as a young academic) asked Professor Ayer what he was most proud of among all his achievements in philosophy. 'Well, I killed off moral philosophy as a subject. That was quite an achievement,' was the reply. So AB, unaware of the full picture, but knowing of at least one other development, said that this was very interesting, because he had come across a book called *Reasons and Persons* by Derek Parfit,[1] which had made quite an impact, with an international conference devoted to it. This brought a dry smile. 'He does seem to have done a good job at reviving the subject.' Only years later did AB learn that in a BBC interview back in 1978 Professor Ayer had said of logical positivism, which was the foundation of his philosophy, 'I suppose the greatest defect...is that nearly all of it was false.'[2]

AB bought the house in which he still lives from Derek Parfit's family in 1984, the same year in which *Reasons and Persons* was published. The book is an analysis of what it means to be a person, and in the light of that how we should live. The first quarter of the book, written in the heyday of Thatcherism in Britain, is a demolition of the self-interest theory. *Theory* here is used in the technical philosophical sense, not of a testable scientific idea, but rather a principle by which a person decides what to do next. The self-interest theory is the idea that a person ought to do whatever is in their own self-interest. To this day you often hear a politician asserting, generally of some foreign policy or negotiation,

---

[1] D. Parfit, *Reasons and Persons* (Clarendon Press, 1984).
[2] 'Logical Positivism and Its Legacy', *Men of Ideas*, A. J. Ayer, interview by B. Magee (BBC, 1978).

that such and such a course is in the national interest, as though that were something to be proud of or somehow justified the approach. What Parfit showed was not so much that the self-interest theory is morally wrong, as that it is logically untenable. He did this partly through game theory, whereby you set up a situation in which what is best for each is in conflict with what is best for all, so that an individual who follows the self-interest theory achieves the worst outcome. An example from everyday life is traffic flow. For many road systems, if each driver seeks to make their own journey as short as possible, the result is that everybody's journey takes longer. The technical way to alleviate this is through a pricing mechanism, such as congestion charging. But pricing mechanisms for public goods cannot avoid moral questions, as becomes apparent in the provision of health care. Parfit goes on to consider what it means to be a person, because how to live your life depends on what it means to be you.

Shall I be the same person tomorrow as I am today? Most people would answer 'Yes.' How then do I balance my self-interest today, say the pleasure of eating pudding, with my self-interest tomorrow, not wanting my clothes to appear to shrink? The worldwide epidemic of obesity shows how hard people find that. The UK government now gives people who retire greater choice in how much of their pension fund to cash in straight away instead of saving it for long-term income. How is that conflict of self-interest to be resolved?

Here is a thought experiment, born out of the invention of the fax machine. This was a machine used to send a facsimile of a written message to another place without needing to send an actual piece of paper by airmail (it has long since been superseded by email). Suppose a machine is invented that is capable of completely digitizing and replicating your body, atom for atom, molecule for molecule, chemical bond for chemical bond, memory for memory. Suppose that instead of all the hassle of security and delays at airports you could step into the machine in London, and allow your whole body to be digitized and the information transmitted to New York where a copy of your body would be 3-D printed.[3] Wouldn't that be more convenient and pleasant than travelling economy class? It would be almost as good as Scotty beaming up Captain Kirk. Oh yes, I meant to tell you, once the copy has been

---

[3] This thought experiment was proposed in Derek Parfit, *Reasons and Persons* (1984); a more recent version using AI to assemble quarks and electrons is in Max Tegmark, *Life 3.0* (Alan Lane, 2017), 225.

verified as accurate and living, you are required to press the destruct button in London. Would the body now in New York be you? Yee-ess, you say, maybe a little hesitantly, but perhaps you could be persuaded. After all, you would have your body, and your memories, and there would be no other you, and you could now do whatever you went to New York to do, just as though you had flown by aeroplane, but much more comfortably and conveniently. Ah, but what if the you back in London hesitated, and lost courage or confidence, as if Scotty had left a version of Captain Kirk behind who had to destroy himself but had second thoughts? What if you did not press the destruct button? Who would then be you? And how should each of you live? What does self-interest mean in this situation? Who has 'your' job under employment law? Who owns 'your' house? What about the outing you planned with your daughter for her birthday? And who is responsible for that hurtful comment you made last week?

Towards the end of the book Parfit introduces Theory X, to replace the self-interest theory. He offers various criteria which Theory X should satisfy, and some properties which Theory X should manifest. But then he confesses that he cannot find Theory X. It is like reading an Agatha Christie mystery only to find that after Hercule Poirot has assembled all the suspects in the drawing room the last few pages are missing. Maybe this is inevitable, perhaps because the search for Theory X involves a subtle self-contradiction. It might be described as a person looking for a method by which they can be a non-person.

The question of how I should live is inseparable from the question of who I am. The concept of *Homo economicus* was widely used following John Stuart Mill's work on political economy, but it is now something of a period piece. *Homo economicus* was supposed always to act rationally in his own best interest, to maximize either his utility as a consumer or his profit as a producer. This is now widely recognized to be a gross simplification. *Identity Economics*[4] recognizes how our behaviour is shaped by our social context. An identity may be one about which we have no choice, such as our race or our age, or it may be one which we choose, such as volunteering for the armed forces. Some identities, such as gender, most people through most of history have simply accepted, but that is changing. The right to choose religious identity is enshrined in Article 18 of the Universal Declaration of Human Rights. Each aspect of

[4] George A. Akerlof and Rachel E. Kranton, *Identity Economics: How Our Identities Shape Our Work, Wages, and Well-Being* (Princeton University Press, 2010).

my identity can have a huge influence on what I choose to believe, and hence on how I choose to live.

Andrew Briggs recounts:

I have vivid memories as a child, aged perhaps 11, of worrying precociously about free will and determinism. If, as seemed to me to be the case, the brain is a fully deterministic machine whose operation is amenable to unlimited scientific investigation and description, what does it mean to be me? Specifically, how can I make choices if everything is determined by the firing of neurons in my brain? Needless to say, the 11-year-old me did not have the benefit of all the thinking by others much better qualified and informed, but that did not stop me from experiencing the force of the question. Years later an atheist Nobel Laureate would write, 'You, your joys and your sorrows, your memories and ambitions, your sense of personal identity and free will, are in fact no more than the behavior of a vast assembly of nerve cells and their associated molecules.'[5] I have found that many other people whom I have told about this respond that they too worried about free will at about that age.

Although I have since read and thought much about free will, I am no nearer to finding a fully satisfactory answer. Those who like labels call me a compatibilist, by which they mean that I consider that free will is compatible with a mechanistic description of the brain. They may even call me a dual-aspect monist, by which they mean the view that the mental and the physical are two aspects of the same thing.[6] Perhaps that is the closest that I can get to a satisfactory position, though satisfactory is too strong a claim. I find it more helpful to think about responsibility, which I find to be a more sharply defined concept than free will. I do not know how free I have been in the various choices that I have made, how much I have been influenced by my parents (probably much more than I realize), my teachers, my pastors, and my friends (though in some cases I can identify strong influences, most of them beneficial). It is tempting, and maybe not entirely inaccurate, to follow the convention of authors in acknowledging friends and colleagues for their helpful advice and input but taking responsibility for the deficiencies. And that is the point. Without being able to quantify the degree of free will in each case, the only way that I can live my life is to take responsibility for the choices that I make and the decisions that I take. Without that, I would not know what it means to be me.

---

[5] F. H. C. Crick, *The Astonishing Hypothesis—The Scientific Search for the Soul* (Simon and Schuster, 1994), 3.

[6] M. A. Jeeves and W. S. Brown, *Neuroscience, Psychology, and Religion: Illusions, Delusions, and Realities about Human Nature* (Templeton Press, 2009).

5.1 Machine intelligence

No field of computer science, perhaps no field of technology, is making faster progress than machine learning (we'll come back to humans in the next section). Almost anything we write about this will become out of date more quickly than any other topic in this book. But let us record a contemporary snapshot.

If a British citizen returns to the UK through Heathrow Terminal 5 with an EU passport (well, perhaps that will become out of date even more quickly), the machine reader will check the passport faster and more reliably than an experienced immigration officer. Facial recognition is something that humans are very good at, but it is an intrinsically difficult task. Different angles, lighting, facial expression, and hair arrangement and spectacles, all make it difficult to compare the face with the passport photograph with the required level of confidence. Three developments have enabled machines to do this task well, in increasing order of importance: (i) faster computer hardware; (ii) better algorithmic software; (iii) larger labelled data sets. The last is by far the most significant. The facial recognition used by the passport machines has been trained on over 50 million faces. A human who saw one face per minute each working day for forty years would not see even a tenth of that number.

Andrew Briggs recounts:

I was privileged to have inspiring maths teachers at school. One of my teachers, actually the head of maths, got us to make a computer that would learn how to play noughts and crosses (in America this is known as tic-tac-toe). We took a number of matchboxes (they were more common in those days) and labelled one for each possible position (there are not many if you allow for symmetry). Into each matchbox we put beads colour-coded for each possible next move, about three beads of each colour. We inserted diagonal dividers into the matchbox drawers, so that by shaking and tilting a matchbox, and then slightly opening the drawer, one bead would appear at the apex formed by the divider. That determined the play for the computer to make at that point in the game. Two sets of matchboxes allowed for either the computer or the human to make the first move. After playing a game, if the computer won it was rewarded by adding a bead of the appropriate colour to each matchbox involved. If it lost, a bead of the appropriate colour was removed from each matchbox involved. Following a draw no changes were made. Noughts and crosses is a rather dull game, since good play on both sides invariably results in a draw, which is why it is mainly played by small children until they work that out and get bored. But it was simple enough for us to make our little matchbox computer. Quite soon the computer became good enough never to lose. I did not know it at the time, but what we were doing is called reinforcement learning.

Every time someone surfs the Internet they are the beneficiary, or the victim, of machine learning. Huge commercial investment has been poured into machine learning for targeted advertising, as any online shopper should appreciate. What they may not appreciate is that machine learning is also used for discriminatory pricing; for items whose price can change (such as airline tickets), machine learning can be used to determine what a given customer would be willing to pay, and therefore price the item accordingly (a kind of automated version of haggling in an oriental bazar). It may be that the wider public will first be persuaded of the power of machine learning when insurance premiums for driver-less cars become cheaper than premiums for cars driven by humans. Actuaries are a hard-nosed profession, and they can be expected to take a rigorous approach to the relative safety of self-driving cars.

The range of activities in which the intelligence of the machine sur-passes that of a human has grown inexorably. Suppose you lay out on a table a range of human mental skills, with those requiring least intelli-gence closest to you and those requiring most intelligence further away. It is not easy to define or measure intelligence, but for the present pur-pose it is sufficient to recognize that solving a hard crossword puzzle requires a higher level of intelligence than adding two plus three. You can then imagine a boundary within which machines can exhibit a higher level of intelligence than humans, and outside which humans exhibit a higher level of intelligence than machines. With time that boundary has moved outwards. Early pocket calculators were better and faster than humans at long division, and at taking the square root of an arbitrary number, but they could not play chess. In 1996 Gary Kasparov played a match of six games against the IBM computer Deep Blue. He won. A computer is not as good as the world champion at playing chess. In 1997 he played a second match against Deep Blue. He lost. Yes it is.

Go is a significantly more difficult game for a machine to play than chess, even though the rules are simpler. Each player has only one kind of piece, a stone. The board consists of 19 × 19 lines, forming a square grid of intersections. Players take it in turn to place a stone of their colour on a vacant intersection. A stone is captured and removed when each immediately adjacent intersection is occupied by an enemy stone. No stone may be placed so as to recreate a former board position (thus avoiding endless circular loops). Two consecutive passes, when neither player can or chooses to make an allowed move, end the game. The player with the most territory wins, counted as all the intersections occupied or surrounded by that player. That's about it.

There are two reasons why Go is harder to play than chess, and much harder for a machine. First, there is a wider choice of moves. At a stage where in a game of chess there might be a choice of 20 moves, in Go there might be a choice of 200 moves. If you are trying to foresee the possible consequences of one of those moves, then you must consider 200 possible responses by your opponent, to each of which you might respond with one of 200 moves, to each of which your opponent might.... The exponential is a very powerful function when it comes to considering all the possibilities. Secondly, whereas in chess there are well-developed ways of estimating which player is in the stronger position (start by adding up the pieces captured, assigning an appropriate value to each, and then consider the areas of board controlled by each player), in Go it is much more difficult to say who is winning at any stage in the game, and hence to calculate the relative benefits of alternative moves. That is why it took much longer for a computer to play Go than chess. What made it possible was that whereas Deep Blue was programmed to win at chess, AlphaGo was programmed to *learn* how to win at Go.[7]

It is this ability to learn that marks out current advances from earlier artificial intelligence.[8] That is why it is possible for machines to be better at facial recognition than humans, and other kinds of recognition too. AlphaGo uses reinforcement learning, the same as we did for our noughts and crosses computer at school. It first learns from a large number of games available on the Web, and learns patterns of play which led to wins. It then tries them out on another version of itself. The computer that won rewards its winning moves, while the other one, as it were, punishes the moves that led to it losing. Like good sparring partners,

[7] D. Silver *et al.*, 'Mastering the Game of Go with Deep Neural Networks and Tree Search', *Nature* **529** (2016), 484–9.

[8] Tegmark, *Life 3.0.*

the two computers allow each other to improve from what the other has learned. A little bit of randomness is added to each move, to keep the computers learning. In 2015 AlphaGo played the European champion Fan Hui. AlphaGo won 5–0. In the game against Lee Sedol the next year AlphaGo won 4–1, losing game 4. The leader of the team that created AlphaGo, Demis Hassabis, estimated that more than 100,000,000 people followed the match, with strong interest in Korea, China, and Japan. One of the AlphaGo team has since announced that AlphaGo has learned from an error which it made following Lee Sedol's move 78 in Game 4, to which losing that game was widely attributed. He assures us that it will not make the same mistake again. In 2017 the program has been playing with even greater success. The most recent version, AlphaGo Zero, was entirely self-taught, without using human data or any guidance beyond the rules of the game.[9] AlphaGo illustrates how far machine learning has progressed since the field of artificial intelligence emerged in the middle of the twentieth century, soon after digital computers were shown to be viable.

In AB's laboratory, researchers are implementing a kind of machine learning known as Bayesian optimization. The Reverend Thomas Bayes was a Presbyterian minister, and the son of a Presbyterian minister, who devised a solution to a problem of inverse probability. His paper on the topic was read to the Royal Society in 1763, shortly after he died. Blaise Pascal and Pierre de Fermat had laid the foundations of probability in the context of gambling. The particular problem which they worked on together was how the winnings should be fairly divided in a game that was interrupted before the end. Their innovative insight was that the answer depended not so much on who had won what so far, but on the possible ways in which the game might have continued. Their thinking led to the concept of an *expectation value* which has become a widely used concept in science and provides the basis for the actuarial calculations used by every insurance company.

Some results in probability seem counterintuitive. Try it yourself: if parents have four children, are they more likely to have two of each sex or three of one and one of the other? You will find the answer at the end of the chapter, but please try the problem yourself first (no cheating allowed).[א]

---

[9] D. Silver *et al.*, 'Mastering the Game of Go without Human Knowledge', *Nature* **550** (2017), 354–9.

Thomas Bayes turned probability the other way round, and asked how you could use fresh observations to reassess the probability that a belief is correct. If you have a jar which you know contains 25 black balls and 75 white balls, you can readily calculate the probability that if you draw out four balls one of them will be black and three will be white. (We say 'readily' in the sense that a high school maths student could figure this out.) Now invert that problem. Suppose you have a jar which with 90% confidence you believe contains 25 black balls and 75 white balls. Then suppose you draw four balls at random and three of them are black and only one is white. In the light of this new evidence, what now is the probability that your original belief is correct? What is the probability that it would be correct to believe that the jar actually contained 75 black balls and 25 white balls? Bayes worked out how to evaluate that.

Bayesian analysis can be used to give reasons for belief where uninformed intuition might be otherwise misleading. Suppose I am concerned about whether I have a rare cancer for which the incidence in the population as a whole for my age and sex is ten in a million. I take a test which gives a positive result for 99% of those who have this cancer, but which also gives a false positive with 5% probability for the same age and sex group. If I test positive, what should I believe about the probability that I have cancer? Many people would guess wrongly: 99%, 95%, or 5%? Each of these answers is incorrect. In Section 11.7 we shall give Bayes' formula, which tells you that the correct answer is 0.02%. This is the probability that I am correct to believe that I have cancer.[10]

Suppose you have a belief about how people will vote in a forthcoming referendum or election. But there is uncertainty in your belief. You can reduce the uncertainty by acquiring more information. The information may be costly to acquire. It might involve conducting an expensive opinion poll. Bayesian optimization is all about choosing what information to acquire next, under constraints of time or cost,

---

[10] The reason that one may find this answer surprising is that many of us have a poor intuition about rare events. In a population of one million people in this example, on average 10 will have the disease, and 999,990 will not. If all take the test, then the ill people will (very likely) get a positive result, and in the healthy group, 50,000 will give a false positive and the rest a correct negative. So, if one has a positive test result and no other information except the general rareness of the disease, then one is far more likely to be among the 50,000 healthy people than to be one of the ten ill people.

that will do the most good in reducing the uncertainty in the belief, and thereby increase the accuracy and confidence. As examples of voting illustrate time and again, 100% accuracy and confidence are unachievable. Nevertheless, the principle of reducing uncertainty by acquiring well-chosen new data is a good one. Rather like Darwinian evolution, once you see it you cannot unsee it. Once you see it, you realize that you have been using Bayesian optimization all your life, quite possibly without knowing it. Your friend tells you something. You believe you have understood correctly, but you would like to be more sure. So you ask a question which will elucidate more information from your friend, to increase your confidence and the accuracy of your belief. Techniques of machine learning are being developed that enable machines to implement this kind of optimization.

Within a few years machine learning will be as widespread in laboratories for deciding what to measure next as computers already are used for controlling apparatus and for recording data. Bayesian optimization is used in science to help choose molecules for specific applications. For Bayesian optimization you need to calculate an acquisition function, which basically tells you how much benefit a particular new data point will yield. A key development has been determining the maximum in the acquisition function much faster than calculating or measuring the data point itself, so that there is a genuine advantage in taking the time out to make the best choice. Bayesian optimization can even optimize how long to spend making the choice; you might call this optimizing the optimizer. In a laboratory the belief might be a mathematical formula describing the behaviour of a device, or it might be a set of parameters for the best performance. AB has set his laboratory the ambitious goal that within five years the machine will be deciding what to measure next to the standard of a second-year graduate student. As we were finishing this book, we received the results of the first experiment in his laboratory in which the machine had decided the choice of measurements. Current graduate students inevitably ask what will become of them, to which they are told that they should get their PhD quickly before they are overtaken by machines!

The inexorable rise in machine learning raises fresh questions about the meaning of intelligence, by asking in what way attributes which we commonly ascribe to humans can also be manifested by machines. Let us examine some of these.

In the opening talk of a conference on artificial intelligence in 2016,[11] the distinguished entrepreneur Hermann Hauser defined intelligence as 'knowing what to do next'. He then added the qualifier 'in pursuit of a goal'. In that sense we have already seen that machines can do this, all the more so if you reckon that the best way to predict the future is to create it. We shall return to creativity later.

Can a machine have imagination? For his PhD Demis Hassabis, who led the AlphaGo team, studied the relationship between imagination and memory.[12] He made comparisons between normal patients and patients who had some impairment to their memory such as hippocampal amnesia. A typical experiment might ask a subject to do the following:

(i) Please conjure up a memory of a beach holiday which you have enjoyed. Please describe it in detail. Psychologists have ways of measuring the sharpness of the memory and the amount of detail recalled.

(ii) Please imagine a beach holiday that you would like to have. Please describe in detail what you are imagining. The same psychological measures are applied.

What the research found was that there is a strong correlation between the ability to recall the past and the ability to imagine the future. This can be applied to machines. AlphaGo was able not only to recall moves that had been played in the current and other games, but also to put forward and evaluate future moves that might be made in that game. Because the evaluation here relies so heavily on the memory of past games, the ability to 'imagine' in this sense is linked to the ability to recall previous moves.

Can a machine exhibit creativity? In move 37 of game 2 against Lee Sedol, AlphaGo made a move that took everyone, including Lee Sedol, by surprise. It was at a stage in the game where a top player might choose between placing his stone on an intersection with line 3 or line 4. Line 3 is likely to give more territory, and since the aim of the game is to acquire territory that is a good place to be. Line 4 is likely to give more influence, and since you can win more territory later by having

[11] 'Machine Superintelligence and Humanity: Rustat Conference Report', Jesus College Cambridge (23 August 2016), available at https://www.jesus.cam.ac.uk/articles/machine-superintelligence-and-humanity-rustat-conference-report.

[12] D. Hassabis, D. Kumaran S. D. Vann, and E. A. Maguire, 'Patients with Hippocampal Amnesia Cannot Imagine New Experiences', *Proc. Natl. Acad. Sci. USA* 104 (2007), 1726–31.

more influence now, that is also a good place to be. There is a parable there for those who have ears to hear. AlphaGo chose an intersection with line 5. If, for a moment, you will allow that an aspect of creativity is doing something which no one else would think of, but which after you have done it everyone else who understands these things can see is brilliant, then that is creativity.

Can a machine have a belief? We have already seen that this is how Bayesian optimization works. The optimization does not start from no prior expectations at all (in the probabilistic sense), but rather adjusts the likelihood estimates by bringing in new information. This is reminiscent of what we mean by belief in the context of human thought, and it is noteworthy that this way of proceeding is beneficial both in automated processes and in the way humans do science. One of the differences between earlier forms of automated science, in which machines were sometimes described as formulating and testing a hypothesis, is that earlier forms started with the computer equivalent of an empty head.[13] It was remarkable how, with no prior knowledge at all, machines were able to deduce equations governing chaotic double pendulums and even to win computer games given only the values of each pixel and the score at any moment.[14] But that was a test of what can be done under a severe limit on the prior information. It is a slow way to do science. It is much faster to give the machine a prior belief, and then allow the machine to refine that through Bayesian optimiza-tion. This is how a human does science. In a typical science laboratory in Oxford every graduate student is required in their first year to write a review of the relevant literature; one of the criteria by which the lit-erature review is assessed is that it should not omit any paper which, if the student had known about it, would have affected how they con-duct their research project. No practising scientist can work without prior beliefs. Max Plank, whose work on quantum theory won him the Nobel Prize in Physics in 1918, wrote, 'Anybody who has been seriously engaged in scientific work of any kind realizes that over the entrance to the gates of the temple of science are written the words: *Ye must have faith.* It is a quality which the scientist cannot dispense with.'[15]

---

[13] D. Waltz and B. G. Buchanan, 'Automating Science', *Science* **324** (2009), 43–4.

[14] V. Mnih et al., 'Human-level control through deep reinforcement learning', *Nature* **518** (2015), 529–33.

[15] M. Planck, *Where Is Science Going?* (AMS Press, 1932).

Can a machine be surprised? Surprise is an essential quality for a scientist. Alexander Fleming had a reputation as a brilliant researcher with an untidy laboratory. In the summer of 1927 he set off on holiday, leaving some cultures of staphylococci on a bench in a corner. When he got back from holiday he noticed that one of the cultures had become contaminated with a fungus, and that the neighbouring cultures of staphylococci were dead. His famous remark was 'That's funny!' His surprise led to the discovery of penicillin. When graduate students plan an experiment, they are trained to articulate what they expect, so that if the outcome is different, they can recognize it and investigate further. The strength of evidence needs to be in proportion to the extent of the surprise. In Bayesian optimization, the machine has a prior belief. When the next data point is acquired, the extent to which it differs from the prior belief can be taken as a measure of surprise. The machine can take that into account in deciding what to measure next.

Can a machine be curious? At one level, curiosity is something shared with all life. Even a bacterium will explore its surrounding to find a higher concentration of fuel. Anyone who has a cat can understand where the idea that cats manifest curiosity comes from (the Briggs' family cat also manifests clairvoyance, to judge from the way that it unerringly sits in whichever armchair Andrew was about to sit in). It seems that a higher level of curiosity is integral to being human. There is a great deal of evidence that our desire to understand the material world is driven by something more than simply wanting to develop technology. This has been averred repeatedly by scientists, and investigation of natural phenomena has time and again flourished in cultures that pursue questions about reality beyond what is immediately visible or tangible to the senses.[16] Whether a machine can be curious may be related to whether a machine can choose a purpose.

What would it mean for a machine to choose a purpose? At present, the hierarchy of goals of a computer are set by the programmer through what is called the utility function. Some engineers refer to this as the loss function, but that term was not dreamt up by a public relations expert. 'This machine makes the loss function more negative!' does not sound catchy to a non-specialist. It is more attractive to a non-specialist to speak of maximizing the utility function. The utility function is just that, what

---

[16] Roger Wagner and Andrew Briggs, *The Penultimate Curiosity: How Science Swims in the Slipstream of Ultimate Questions* (Oxford University Press, 2016).

the programmer decides the machine should maximize. (There is another parable here: what are you seeking to maximize in your life?) It might be the accuracy with which the properties of a sample are measured, or it might be optimizing those properties against some criterion. In another walk of life the task might be to analyse a company's accounts, with the utility function being to minimize the tax to be paid, or to make the tax bill a fair and reasonable reflection of the company's turnover. There will often be multiple aims, which might be in conflict: maximizing the reliability of an electronic component while minimizing its cost. A self-driving car has a hierarchy of goals: get me from here to there; get me there in the shortest time; obey traffic laws; do not kill or injure anyone; if an accident is unavoidable, cause minimum deaths and minimum injuries. How are these to be chosen? If you were manufacturing a self-driving car, would you make it a selling point that if an accident cannot be avoided, then a machine-learning equivalent of an air bag would prioritize not injuring the occupants over other road users?

As the boundary separating what the machine is better at from what humans are better at moves relentlessly outwards, might the day come when machines set their own hierarchy of goals?[17] Will they be able to reset their initial utility function? If so, on the basis of what values will machines make such decisions? You might suggest that they should be given human values.[18] Whose human values? Those of Daesh? Well, yours and mine of course. Except…I am not so sure about yours. Perhaps it would be more like parents bringing up a child: they may communicate their values by word and by deed, but there is no guarantee how their adoption will navigate the teenage transition. Whatever values are somehow imparted to the machine, what would it mean for the machine to be responsible for what it does?[19]

Let us now summarize the thrust of the discussion in this section. We have surveyed various aspects of machine intelligence, approaching it largely from a practical point of view. We have not tackled the philosophical subtleties that surround issues of consciousness, but simply explored the fact that machine intelligence is making rapid advances

[17] Nick Bostrom, *Superintelligence: Paths, Dangers, Strategies* (Oxford University Press, 2014).

[18] Murray Shanahan, *The Technological Singularity* (MIT Press, 2015).

[19] G. A. D. Briggs and D. Potgieter, 'Machine Learning and the Questions It Raises', in *From Matter to Life*, ed. S. I. Walker, P. C. W. Davies, G. F. R. Ellis, 468–86 (Cambridge University Press, 2017).

and is showing signs of some of the features associated with thought. One should not exaggerate this, since, after all, even a humble earthworm shows a range of problem-solving skills as it avoids toxins and seeks moisture and repairs its body; the sophistication of the processes going on in its cells easily surpasses those of a modern supercomputer. Nevertheless, it is not hard to imagine that, sooner or later, a machine based on a computer will show every appearance of having hopes and desires. Like HAL in *2001: A Space Odyssey*, it might even seem to develop a mind of its own. What should be our attitude to such thinking machines? Perhaps our general attitude should be to be ready to err on the side of generosity. That is, one would rather be duped into acting well towards a machine which, in fact, had no pain or sadness, than ignorantly hurt a machine which could experience pain. The Golden Rule can be generalized as the notion of making an effort to discover or guess what another's hopes or goals might be, and to act to facilitate those goals, in so far as this can be done while also taking into account others whose interests may conflict with the first. In the future our descendants may have to become more and more sensitive to what the possibilities may be. If an autonomous computing device were one day to achieve cognitive and empathic abilities like our own, then perhaps that device might seek and express an accordingly deep relation to God, with rights and obligations like ours.

5.2 How did humans become responsible?

Machine intelligence is not subject to the same hardware constraints as human intelligence. The size of the human brain is subject to the demand for fuel and the constraints of the birth canal. A larger brain relative to the mother's pelvis leads at best to pain and at worst to death in childbirth. The difficulty is in part overcome by more of the growth taking place after birth. The human brain at birth is a much smaller proportion of the adult brain size than other primates, with the rapid rate of prenatal growth continuing for the first year or so after birth.[20] This growth carries with it high metabolic and nutritional demands. The demand for fuel for the brain continues for life.

[20] W. R. Leonard, J. J. Snodgrass, and M. L. Robertson, 'Comparative and Evolutionary Perspectives on Human Brain Growth', in *Human Growth and Development*, 2nd edn, ed. Noël Cameron and Barry Bogin, 397–413 (Academic Press, 2012).

A major constraint on neural design is energy consumption.[21] Why do animals have brains? One way of thinking about the transition from inanimate matter to life is in terms of information management.[22] This can give a rather different perspective from the definition of life in terms of reproduction which we may have learned at school.[23] The simplest bacteria exhibit information management. Detailed measurements of general intelligence show that various components of intelligence are present in a range of mammals (some of the most detailed studies have made comparisons with rodents and with other primates[24]). Birds, those wonderful survivors of the climate change that wiped out all the other dinosaurs 66 million years ago, show striking powers of problem solving. If you put some food in a miniature bucket at the bottom of a glass cylinder, too deep for the bird to reach, and leave a wire nearby, a crow will use its beak to bend a hook in the wire, which it can then use to extract the bucket. More remarkable still, if the bucket is floating on water in the cylinder, still out of reach, and you leave some rubber and some polystyrene objects nearby, the crow will experiment and discover that the polystyrene objects float and do no good, but the rubber objects raise the level of the water and thereby bring the food within reach, just like the thirsty crow in Aesop's fable.[25] Only the most precocious four-year-old human will do better!

It should not surprise us that human intelligence uses hardware that we share with other species. Although individual steps in evolution may be random, the overall direction is constrained by the way the world is. Suppose there is an advantage in being able to see, so that an increasingly sophisticated seeing organ would evolve from some primitive cell that has sensitivity to light. There are only a limited number of ways that you can make an eye. One is the compound eye, which is

[21] P. Sterling and S. Laughlin, *Principles of Neural Design* (MIT Press, 2015).

[22] P. Nurse, 'Life, Logic and Information', *Nature* **454** (2008), 424–6.

[23] S. Imari Walker and P. C. W. Davies, 'The Algorithmic Origins of Life', *J. R. Soc. Interface* **10** (2013), 20120869.

[24] J. M. Burkart, M. N. Schubiger and C. P. van Schaik, 'The Evolution of General Intelligence', *Behav. Brain Sci.* **40** (2017), e195.

[25] S. A. Jelbert et al., 'Using the Aesop's Fable Paradigm to Investigate Causal Understanding of Water Displacement by New Caledonian Crows', *PLoS One* **9** (2014), e92895. It has been questioned to what extent these experiments betoken causal understanding; S. Ghirlanda and J. Lind, '"Aesop's Fable" Experiments Demonstrate Trial-and-Error Learning in Birds, But No Causal Understanding', *Animal Behaviour* **123** (2017), 239–47.

what insects use. Another is the camera eye, which is what we use. So do octopuses. If you trace back the common evolutionary pathway of humans and octopuses, the divergence occurred long before the development of the camera eye (and there are some subtle differences, such as the retina being beneath another layer of non-photogenic nerve cells in vertebrates but on the surface in octopuses). The reason they both work the same way is that each of them is optimized within the same laws of physics. The technical term for this is convergence.[26] If greater intelligence carries an evolutionary advantage, then we should expect this to be shared between species.

Greater brainpower comes at a cost. The human brain runs at about 20 W, which converts to a daily energy cost of over 400 kcal. This is nearly 25% of the resting metabolic rate of the human body. For comparison, our skeletal muscles typically require about 500 kcal of fuel per day, though an athlete requires more. There must have been a strong driver for the human brain to get so big. There is growing evidence that practical challenges such as finding, catching or processing food may have been even more significant than the benefits of social interaction, whether for hunting or for warfare.[27] As we mentioned in Chapter 3, most of the expansion of the human frontal lobes was complete by about 200,000 years ago, long before the appearance of many of the uses to which we now put our brains. It is very remarkable that the brain whose evolution had already happened that long ago should prove to be so suitable for tasks that only became possible much more recently. Appealing though it would be to think that the research achievements of the authors have contributed to attracting a mate or to reproductive success, we know of no evidence to support such a conjecture. It remains an open question why the human brain is so suitable for activities so far removed from its evolutionary development.

5.3 The distinctive nature of human existence

If you were writing, let's say about three thousand years ago, or maybe a little less, how would you set about describing the distinctiveness of being a human? You are writing for people who know nothing of modern

[26] Simon Conway Morris, *The Runes of Evolution: How the Universe Became Self-Aware* (Templeton Press, 2015).
[27] M. González-Forero and A. Gardner, 'Inference of ecological and social drivers of human brain-size evolution', *Nature* **557** (2018), 554–7.

genetics, or processes of natural selection, or even evolution, and your concern is not about any of those things but about what it means to be human; a person; a 'me'. You are writing for a storytelling culture, and so you would probably put it in the form of a story. Let's say you set it in a garden. The garden is pleasant, but it is also designed for character formation, and so there is work to do, and also the possibility for a hard moral choice. You want to convey that humans need social interactions (for the same reason that solitary confinement is a severe punishment), and so you try the literary thought experiment of having one solitary man and letting him encounter animals and name them. Animals can be useful and they can be good company.[28] But ultimately no animals, not even a cat, are fully satisfactory as partners in work and companions in life. Humans need humans. An enriching component of human relationships is sex. So the supreme gift to the solitary man in our story is companionship with an equal who is both like and unlike; a woman. It is hardly a complete account, but it is a good start. Oh, and there is one other aspect. They should be free of the shame which lies at the root of so much psychological disorder.[29]

As far as it goes, would you regard such an account as complete? If not, what would you add next? The alert reader will already see where this is going. To be fully human, you need the additional ingredient of responsibility.[30] You need to be accountable for your actions. Within the worldview of that time, it would have been widely understood that ultimate accountability is to God. So you let the humans face the moral choice that you had built into the story, and see how they get on. By all means introduce an element of disinformation that will make the moral choice harder. And when it all goes horribly wrong, and the humans realize that they are accountable for their actions, let them blame one another and shirk responsibility, and let them now experience shame and its amelioration.

---

[28] A task for which even cats have a use is keeping rats at bay. In his work for lepers in India, Paul Brand observed how having a cat could protect leprosy patients who had lost all feeling in their feet from being eaten by rats when they were asleep. Dorothy Clarke Wilson, *Ten fingers for God: The Life and Work of Dr. Paul Brand* (Paul Brand Publishing, 1996).

[29] D. F. Allen, *Shame: The Human Nemesis* (Eleuthera Publications, 2010).

[30] A longer discussion of the two trees and their meaning is given in A. M. Steane, *Faithful to Science* (Oxford University Press, 2010).

This is, in a brief sketch, how questions of human distinctiveness and human responsibility were addressed and discussed in early Jewish thought. As an early articulation that to be human means to be accountable, the story of Adam and Eve is unsurpassed.

By the time of Jesus, Jewish scholars were already treating the Adam and Eve story as figurative literature rather than as history.[31] Among the letters of the early Jewish convert Paul, the longest is a letter to his friends and other followers of Jesus in Rome. In the chapter which introduces his core argument, he explains the significance of the death of Jesus. It is for good reason that the logo of Christianity is a cross. Paul wrestles with complementary descriptions to convey the difference that Jesus' death makes. Because of it we are justified, he writes, using legal language. We are saved, in the language of rescue. We are at peace with God, in the language of well-being. We are reconciled, in the language of the family. And the fact that Jesus died for people in advance of the restoration of the relationship is proof of God's love.

Everything that Paul says crucially depends on the death of Jesus as a historical fact, something which few then or since doubt. He then makes a contrast between the detrimental effects of Adam's failings and the beneficial effects of Christ's achievements. Some Christian believers have expressed concern that if you avoid a literal interpretation of the Garden of Eden, then you would undermine what Paul writes about the fact and significance of the death and life of Jesus Christ. But the observation that people have gone astray in their thoughts and actions throughout recorded history is not in question, and it is to this that Paul is alluding. His main purpose is to illustrate the universal application of what Christ achieved. Just as the damage and failure of sin spread to all the human family, now the benefits won by Christ can spread to all who embrace them. That is the point of the comparison Paul draws, and neither the damage nor the benefits are denied by acknowledging that the literary genre of Genesis is highly stylized. Paul's argument turns on the dissimilarity between Adam and Christ in ways that he recounts.[32] The bad effects of human failing are ultimately to be fully

---

[31] E. C. Lucas *et al.*, 'The Bible, Science and Human Origins', *Science and Christian Belief* 28 (2016), 74–99.

[32] C. E. B. Cranfield, *A Critical and Exegetical Commentary on the Epistle to the Romans: Introduction and Commentary on Romans I–VIII*, vol. 1 (International Critical Commentary), p. 288 (T. & T. Clark, 1975).

and generously overcome by the quality of life which God offers through Jesus Christ.

5.4 My genes made me do it

Andrew Briggs' father tells a story from his student days at King's College, Cambridge. With a few other choral scholars and the chaplain he was on holiday in Cornwall. A crowd had gathered on the beach around a body, which a doctor was seeking to resuscitate. It transpired that a boy's father had taken two boys swimming in the sea. One of the boys was his son, the other a friend of his son. In a ghastly moment he realized that the boys were being swept out to sea by the tide. He took the conscious decision to go first to the help of his son's friend, whom he saved, but at the expense of the life of his own son, who drowned. How is one to account for such a choice? To what extent can such behaviour be explained by processes of evolutionary selection?

Evolutionary biology is familiar with the issue and has some relevant things to say. The process of natural selection does not optimize the propagation of one particular instance of a gene. It optimizes the general increase of examples of that gene throughout the population. Charles Darwin wrestled with the question, and even though he was working without the benefits of subsequent discoveries about genetics, he nevertheless articulated an early concept of kin selection. This is the notion that natural selection will favour behaviour that promotes the interests of others having the same genetic material as oneself. The Oxford biologist J. B. S. Haldane later worked this out in more detail. To illustrate the notion, he proposed that it would be natural to expect our instincts to encourage one to give up one's life to save two brothers, since they each share half one's genes, or four nephews or eight cousins. Of course, this does not address the question whether this is the whole of what motivates human behaviour. It mainly stands to show that self-denying behaviour is not in itself mysterious in Darwinian evolutionary terms, but rather to be expected.

More recent work in animal cooperation has drawn on the mathematics of game theory. A simple application of game theory is to public goods problems in which everyone enjoys the benefits whether or not they contribute to the cost. Suppose that I propose to create a park for the local community. I invite my neighbours to contribute, but I make it clear that everyone will gain equal access to the park, regardless of how much money they put in. For simplicity, let us assume that the neighbourhood is socioeconomically homogeneous, so that everyone has similar resources from which to contribute. This can be modelled as a game in which four players each start with 100 euros. They are invited to put money into the pot, which the banker will then double and distribute equally to all four players. If all four players put all their money in, the banker will double that and distribute 800 euros equally, so that each player ends up with 200 euros. A good outcome, you might think. But now suppose that one player chooses to put in only 50 euros. The banker doubles the resulting 350 euros and distributes a 175 euros to each player. They are all worse off than they would have been, except for one player who now has 225 euros. This is an example of a game that is set up so that what is best for each is in conflict with what is best for all. It explains, for example, why it is so hard to get international agreement on carbon emissions, because each nation benefits from the effort made by everyone else in a way that may be only weakly dependent on the cost born by the individual nation.

In a scenario called *the prisoner's dilemma*, two prisoners have been arrested on suspicion of a crime, and are being kept in separate cells from which they are unable to communicate with each other. Each is given the opportunity to admit to the crime or to deny any part in it. If both admit to the crime, they each get a lenient sentence of one year in jail. If they both deny, they each get a more severe sentence of three years in jail. If one admits and the other denies, then the one who admits gets a sentence of four years, and the one who denies is released without penalty. What should each of them do? The best overall outcome is that they both admit to the crime, since then they serve a total of two person-years in jail. But if one of them reasons that the other will take this course of action, then he might calculate that by denying he will avoid any jail sentence at all, albeit at the expense of the other serving four years instead of one. But then the first prisoner might reason that if the second will behave like that, he would reduce his own sentence from four years to three by also denying. In fact, whatever the other prisoner does, each prisoner will reduce his own sentence by denying. This is the really central point here. In the prisoner's dilemma, the action of self-interest seems to be clear: deny. But this very action, adopted by two supposedly rational prisoners, achieves the worst outcome for both of them: a total of six person-years in jail. Derek Parfit used the prisoner's dilemma in his logical demolition of self-interest theory.

Games such as the public goods problem have been extensively used in experiments with human subjects. Changes in behaviour can be tracked over successive rounds. Different refinements can be introduced, such as the ability of one player to fine another whom he perceives not to be contributing fairly. It is found that players are willing to do that even when the rules incur a cost to themselves for imposing a fine. A whole field of experimental economics has grown up using such protocols.

Game theory can be applied to try to understand human altruism. It is helpful here to distinguish altruism, as a conscious choice of behaviour made by a responsible agent, from cooperation, which can be observed in a wide range of animals from insects to dogs. Some studies have not made this distinction, and in most of those, altruism is simply an anthropomorphic metaphor for cooperation. Cooperation can be observed and measured. Altruism requires or at least implies some knowledge of the motivation of the agent, and the reason why they make the choice that they do.

At least five different mechanisms can be identified for cooperation between sentient animals:[33]

(i) Direct reciprocity is effective in repeated encounters between individuals: you scratch my back and I'll scratch yours.

(ii) Indirect reciprocity works when communication is sufficient that a person's reputation spreads through the Twittersphere, especially if social norms have been established.

(iii) Spatial selection can affect who cooperates with whom, and how the boundaries of an advantageous subgroup change as a result of actual or perceived cooperation.

(iv) Multilevel selection can give selective advantages to groups, as Darwin himself recognized, since a tribe whose members were willing to act sacrificially for the common good might be victorious over other tribes.

(v) Kin selection is the mechanism identified earlier, and can be quantified through a genetic relatedness parameter.

Let us suppose that this kind of mathematical game theory can account for all of the cooperation that we see in the animal kingdom apart from those animals that show evidence of moral reasoning,

---

[33] M. A. Nowak, 'Evolving Cooperation', *J. Theor. Biol.* **299** (2012) 1–8.

namely humans. Maybe it already can, or maybe it will one day. Now let us apply that to human behavior. Let us put the kind of behaviour that we would predict using the very best game theory alongside the kind of behaviour that we might associate with religious motivation. It might be the Mosaic law, or Jesus' Sermon on the Mount, or Paul's 'fruit of the Spirit', or something equivalent from another religion or from philosophical humanism. It would be reasonable to ask, 'Wherein do these (i.e. game theory and religious motivation) exactly coincide, and are there cases where we find a difference so great that we should attribute the religiously motivated altruism to something else?'

Here we move into territory of human choices and responsibility, but we have to face it. For each of us, it is integral to our experience of living as a person that we are able to choose to act in a way that overcomes the intellectual grip of determinism, whether from bottom-up causes such as neurons firing or from top-down constraints of mathematical game theory. At the point where our actions are 'up to me', there is a transition from cooperation as a behaviour that can be observed to altruism as a choice that each of us makes. We see that on a large canvass from the life of Mother Teresa, who gave herself for the poor of Calcutta, to the life of Jean Vanier, who gave himself for the mentally differently abled. Such people demonstrate an altruism that it may perhaps be possible to analyse in Darwinian terms, but what if the genetic processes furnish not the actual choices we make, but rather the capacity to make a choice? Our genetics may predispose us one way or another, but what if that which the biological hardware supports is a 'me' that is able to shape my own selfhood by deciding whether to be selfless?

As with all discussions of free will, and the relation of the mind to the body, we are not invoking a god-of-the-gaps approach.[34] The human choice to act altruistically is not a gap in our neuroscience or our game theory that until now science has not fully explained. There is no aspect of the material world in which we are embodied that is not open to those or any other scientific methodologies. Rather, the conceptual transition from cooperation to altruism is one that in principle lies outside the scope of science. For the Christian believer this reaches its zenith in the life of Jesus. Professor Sarah Coakley concluded her final 2012 Gifford Lecture thus:

---

[34] Sarah Coakley, *Gifford Lectures* (Oxford University Press, and Eerdmans, 2017), Lecture 5.

But when we get to Jesus' sacrifice, finally, we meet the arena of 'super-normality', as I've termed it in these lectures. We meet here not the physical power to give birth in pain, nor the evolutionary capacity to get group good out of individual loss, nor yet the Girardian desire to purge violence and so re-establish order (though arguably that was what was unleashed *against* him). No, what we get is an altruism beyond calculation, yet one that—far from being excessive in violence—is, as I've argued in these lectures, excessive in its rationality. It is, finally, superrational, because it does what only needs to be done once, and divinely, in the order of creation, and thereafter entered into by participation in the resurrection that it leads to.[35]

For Jesus, as for us, the ultimate altruism results from a conscious choice to bring benefit to others at potentially unlimited cost to oneself.

5.5 In the image of God

Let us continue our fanciful exercise of how we would explain the creation to earlier generations, maybe 100 years or more after our previous audience. Remember, at the time no one knew that our universe was 13.8 billion years old, or how old our Earth was, or when humans appeared. They did not have the intellectual or cultural framework of quantum theory or general relativity. But they had stories. They lived in a time that loved stories.

One of the creation stories that had wide currency was discovered in 1849 in a ruined library in what is now Mosul in Iraq, the ancient city of Nineveh, and published by George Smith in 1876. It is generally known by the first two words in Akkadian, *Enûma Eliš*. The poem begins,[36] 'When the heavens above did not exist, and the earth beneath had not come into being—... before meadow-land had coalesced and reed-bed was to be found—'. The poem describes how the teenage children gods had rowdy parties with the music turned up much too loud. Apsû and Tiamat, the father and mother and grandparents of all these gods, were fed up with this. Apsû was minded to adopt a nuclear option, but Tiamat was more soft-hearted and wanted to adopt a more conciliatory kind of discipline. Following the death of her husband, Tiamat was persuaded to take a stronger stand. One of the trouble-makers, Marduk, is

---

[35] Coakley, *Gifford Lectures*, Lecture 6.
[36] W. G. Lambert, *Babylonian Creation Myths* (Eisenbrauns, 2013).

eventually commissioned by his father, Ea (there is a plentiful supply of these gods, all interrelated), to tackle Tiamat, who by now had remarried. The others egg Marduk on to cut Tiamat's throat. They meet. Another god, Bel, lets loose an evil wind in Tiamat's face. She opened her mouth and was then unable to close her lips again. The wind distended her belly. Marduk took his chance and let fly an arrow into her mouth. He tied her up and killed her. Her followers scattered, and Marduk slaughtered them. Marduk then returned to the corpse of Tiamat and split her in two like a dried fish. 'One half of her he set up and stretched out as the heavens.'[37] The other half he presumably set down as the Earth.

Even if the people you want to explain things to have not read *Enûma Eliš*, since books were not readily available then and few people could read anyway, they may have been exposed to a culture in which this story and others like it were well known. If some of your intended audience were Jews who had experienced exile in Babylon, then they would have been familiar with Marduk as a prominent Babylonian god who was supposed to inhabit a star. In an intellectual climate which assumed that whichever side won had the superior gods, however much their prophets had told them that was not the case, it would have been natural to be in awe of the Babylonian gods. Against such a cultural background, if you want to explain that all that is a load of hooey and it wasn't like that at all, it would be hard to do better than the opening of Genesis, בְּרֵאשִׁית בָּרָא אֱלֹהִים אֵת הַשָּׁמַיִם וְאֵת אָרֶץ, 'In the beginning, God created the heavens and the earth.' The simplicity by contrast with *Enûma Eliš* is majestic.

We have some descriptions from the Ancient Near East of temple architecture.[38] A good architect starts not with the materials of the building or even with its design, but its purpose. What is it for? These early architectural descriptions are much more interested in function than in form. To engage with that kind of interest, you might choose to introduce the three functional systems that are crucial for human life: time (the pattern of day/night); geography and weather (separation of land and sea, rainfall); and food (photosynthetic life as the basis of the food chain). You might then introduce the functionaries involved

---

[37] Lambert, *Babylonian Creation Myths*, 116.
[38] J. H. Walton, *Genesis 1 as Ancient Cosmology* (Eisenbrauns, 2011); an account for a more general readership is available as *The Lost World of Genesis One* (InterVarsity, 2009).

in these systems. You could evaluate whether each of these is fit for purpose, which is the meaning of the Hebrew word טוֹב (*tôb*; in this context, 'fit for purpose' is a better translation than simply 'good'[39]). You might even exploit the temple metaphor to give a hint as to what the creation is for. Unlike a modern church building, which is designed to accommodate the congregation, a temple in those days was literally God's house. If previous or neighbouring cultures thought that God was distant, inhabiting stars or mountains, the temple explicitly provided a place for God living among his people. For Jews that came to be in Jerusalem. If your hearers were exceptionally alert, they would see in this an indication that the created world is for God to be here with his people, not like a deistic absentee landlord who created the world and then had nothing more to do with it, but rather creating the world in order to be engaged with its inhabitants.[40]

You might also wish to settle once and for all which god was supposed to have created the world. Remember that, in view of their military defeat, people had a lingering doubt whether the God known to the Jews was really so great as the gods of the Babylonians, the superpower of the time. So you could let slip a great throwaway line about where Marduk was supposed to live, 'and he made the stars too.' More importantly, you could imply a kind of equation: let X = that One, God, whom you have heard about through your family and national history, and whom you will read about in the remaining pages of the Bible when it becomes available, and whom perhaps you know through your own experience; let Y = that which is responsible for there being something rather than nothing, with the regularity that you can observe and document so that you can make sense of it. X = Y.

This 'X = Y' is, in broad brush, the aim of the account, the item it is aiming to present.

You might use the similarity of your account with temple architecture to make a further point, one which in the divisions of the Bible appears in Chapter 2 of Genesis. The purpose of the temple lies not in its construction but in its use once it is completed.[41] What was the world for? What did the creator do in your account after he has finished? If you use the word *rest*, you would risk being misunderstood. Your

[39] Lucas *et al.*, 'The Bible, Science and Human Origins'.
[40] Isaiah 45:18, 'God…made the earth…as a place to be lived in.'
[41] Walton, *Genesis 1*.

readers might think it meant that the creator was exhausted and needed to put his feet up and have a cup of tea. Classical commentaries on the creation story in Genesis were often given the Greek title *Hexameron*, referring to the six days of creation. But that misses the whole point of the seventh day. You might be better to say that after all the work that he had done, God enjoyed quality time. This is what the world is for. Not 'rest' as in no activity, but 'rest' as in enjoyment, conviviality: life in celebratory mode.

The 'day of rest' took on a new significance when it moved from the seventh day to the first day of the week, but that is another story.

How would you describe the place of men and women in all this? A convenient metaphor that would come to mind is provided by צֶלֶם. This is often translated as 'image', but the force of the metaphor is better conveyed by 'statue'.[42] In the Ancient Near East rulers would put a statue of themselves in a part of their empire which they seldom visited as a reminder of who was in charge, rather as coins in Britain carry an image of the sovereign. You might therefore, in your account, let the creator decide to make humankind as his image, to indicate that they are intended to be his representative on earth, creatively ordering it on his behalf, and accountable to him. The expression *image of God*, or its Latin form *Imago Dei*, is widely used to denote our human uniqueness. Sometimes it is used to convey the similarity of humans to God, with capacity for reason, creativity, and love. But at the heart of the image metaphor is the concept of responsibility. To be human means to be responsible.

---

[42] Lucas *et al.*, 'The Bible, Science and Human Origins'.

ℵ We said no cheating!...In a family with four children there is a probability of ¼ of one boy and three girls, ⅜ of two boys and two girls, ¼ of three boys and one girl. So the likelihood of two of each sex is less than the likelihood of three of one and one of the other.

6

The two Tabors

Andrew Briggs gives a personal account

Two Tabors had a big influence on me when I was a graduate student at Cambridge. The first was my professor for my PhD. The second was the location of a transforming experience.

Professor David Tabor FRS was the founding professor of what, by the time I went to the Cavendish Laboratory to undertake research for my doctorate, was known as physics and chemistry of solids. His genius was to be able to tackle problems that were important but complex and elucidate underpinning mechanisms that were amenable to simple description and analysis. He was interested in what happens when two surfaces rub together. He coined the term tribophysics, though the field that he created is now generally known as tribology. As a schoolboy I had learned the two laws of friction first articulated by Guillaume Amontons (1663–1705): (i) the frictional force is proportional to the applied load; and (ii) the frictional force is independent of the apparent area of contact. David Tabor had shown how these laws arise naturally from the roughness of surfaces—he once visualized the contact between two surfaces as being like turning the Alps over and placing them on the Himalayas. The true contact area is the applied load divided by the yield strength of the materials, and is generally only a tiny fraction of the apparent contact area. The friction force is the true contact area multiplied by the shear strength. Thus, the coefficient of friction is the ratio of the shear strength to the yield strength. As with any theory you can introduce refinements, but this elegant picture proves to be extremely robust, and applicable to a remarkable range of metals and other materials.

David Tabor set me a problem at the start of my PhD, which was to understand the friction of rubber in terms of its elastic properties. The answer came to me when watching *Der Freischütz* on Boxing Day in the third year of my PhD. This is a German opera by Carl Maria von Weber about a junior forester who needs to win a shooting competition in order to marry the boss's daughter, which I suppose shows you what kind of upbringing I had! I had long been puzzling over the fact that what seemed to me to be the key physical phenomenon, namely the adhering and subsequent separating of the two surfaces, came into every equation which I wrote down, and yet seemed to make little difference to the experimental measurements. I realized that I had to put the quantities which depended on this into both the numerator and the denominator of my key formula, which meant that they cancelled out. This insight enabled me to account for all my measurements in a single equation, which one of my thesis examiners subsequently dubbed the Briggs equation.

From David Tabor I learned habits of rigorous experimental science, and meticulous analysis of the results. He taught me, and my fellow graduate students, to be our own severest critics. The papers we wrote about our results had to stand up to scrutiny by him, and in time we learned to scrutinize them ourselves to the same standard. It was not acceptable to submit a paper containing vague arguments to an international journal and let the peer reviewers identify its deficiencies. We had to make the presentation of the data and the clarity of the conclusions so robust that they would survive even the strictest external review. That is a transferable skill.

David Tabor was a devout Jew, and also a gifted linguist. Every Saturday would find him in the synagogue studying the Hebrew Scriptures—in Hebrew—and discussing them with his fellow believers. We often talked about the Bible together. He understood the nature of spiritual commitment, and when I found myself torn between a career as a scientist and a career as an ordained Anglican priest, he was caring and insightful in helping me to think it through.

Mount Tabor is situated at the eastern end of the Jezreel Valley in Galilee. It is often identified with the Mount of Transfiguration where Jesus took Peter, James, and John for a memorable experience which included a change in his appearance and an instruction to listen to Jesus because God is delighted with him. Peter, who was not always one to think before speaking, suggested building a monument to commemorate what they had seen and heard. This was not to be and we may never know whether these events really did occur on Mount Tabor, but it is as good a candidate site as any.

During one summer vacation while I was a student at Cambridge I was privileged to win a scholarship for a one-month course on the geography and archaeology of Israel, based at St George's College in Jerusalem. This was my first visit to Israel, and I was fascinated to see for myself the settings of so many events recorded in the Bible. Some, like the Sermon on the Mount, could not have left any visible trace, and so any identification with a particular place is a mixture of tradition and speculation. But in plenty of other cases there are strong geographical and archaeological indicators that enable the place to be identified with a high degree of certainty. I particularly love ancient waterworks, because these can stay put for millennia with little doubt about their identity. I have more than once walked through the tunnel which was constructed in time for the defence of Jerusalem in the reign of King Hezekiah in 701 BC, under threat of an impending siege by the Assyrian army led by Sennacherib, though some evidence suggests that it already existed by then. The tunnel connects two points 335 m apart, although because of its curves it is actually nearly 200 m longer. Their height accuracy was greater than their directional accuracy, with a difference of 0.3 m between the ends, corresponding to a slope of 0.6%. The work proceeded from both ends, with big wiggles where they met in the middle, indicated by an inscription and a change in the direction of the arcs left by the tools used for digging the tunnel. Although we can keep learning more about the history, a tunnel like that does not move, and we can have a high level of confidence that I was walking down a tunnel that was there throughout all the events recorded in the Bible since it was made.

The course which I attended included two nights staying in a monastery at the top of Mount Tabor, with the day between free for personal reflection. I spent much of that day looking out over the Jezreel Valley and thinking about what I had been experiencing since arriving in Israel. Up till then it had been all too easy for me to keep separate the stories which I read in the Bible (I still have the leather-bound King James Version which my godmother gave me at my baptism) and the material world in which I live out my life, and which I study through my science. At the top of Mount Tabor I came to see with a fresh intellectual clarity that they cannot be so easily separated. The events recorded in the Bible took place at particular coordinates in time and space. Although I was there at a different time, I was occupying some of the same spatial coordinates, relative to the geographic features of planet Earth. Where I had walked and sat in Galilee and in Jerusalem was where Jesus had walked and sat. A separation between the spiritual significance of Jesus and the material world of our existence is not possible—they are connected, and they are meant to be connected.

Perhaps this should not have been such a big surprise as it was. I was born into a Christian family. My maternal grandfather was a successful businessman, from whom I learned a lifelong habit of giving away the first tithe of my income. My paternal grandfather was, by the time I was born, a canon in residence of Worcester Cathedral. This arrangement had much to commend it. There were four canons, with duties for one. They took it in turns to do duty in the cathedral for three months, and then devote nine months to their own study and writing. My grandfather produced a prolific output of readings, hymns, and prayers for use in schools, especially in the daily assemblies which are still part of English law. The hymns he wrote are in use today even more than in his lifetime. As children we grew up with church on Sunday and Bible stories at bedtime. We also grew up with science. My father, as a musician and classicist, spent more time making things than analysing them scientifically—he even built a railway for us in the garden of the house in which I was born. My mother was a mathematician, and from her and her father I developed an early fascination with how things work. Alongside mechanical devices and electrical circuits I learned about evolution and natural selection. All of this seemed to me then, as it does now, to be integral to a faith in God Who creates 'all things bright and beautiful, all creatures great and small.'

The first intellectual challenge to this came not from anything specifically religious, but from worrying about free will and determinism. The place was Sheringham in Norfolk, where as a family we used to take our seaside holiday each summer. I would walk the cliffs wondering how, if the workings of my brain followed scientific principles, I could be free to make choices that were not fully determined by neurological mechanisms. It coincided with wondering who I would be if my parents had not married. On D-Day my father, as part of the Royal Army Medical Corps, crossed three times to Normandy, bringing back wounded soldiers of all nationalities. He was about to cross a fourth time when he received orders to stay behind. His vessel was hit and all lives were lost. Who would I be if he had crossed a fourth time? What does it now mean for me to experience the world through my eyes and ears and touch, and with my memories? I did not solve either problem then, and I have not solved them since, but I have read more. As has often happened in my life, I find that I can carry on with my life and my faith with fundamental questions unresolved—without losing the curiosity to continue thinking about them.

In due course I studied physics at Oxford, and was awarded a fully funded graduate studentship which could be taken up at any UK university. Top of my list was the Cavendish Laboratory at Cambridge, and I was delighted when David Tabor accepted me to undertake research with him. The Cavendish Laboratory had been completed in 1874, under the direction of the first Cavendish Professor, James Clerk Maxwell. Maxwell was an extraordinary scientist. He wrote his first paper, on the mathematics of multifocal ovals, when he was still a schoolboy. The contents were presented at a meeting of the Royal Society of Edinburgh by a friend of his father, since James himself was far too young. His curiosity as a child was boundless. 'What's the go o' it?' he would repeatedly ask his patient parents (and anyone else who would listen). He contributed experimentally to studies of colour vision and photography, visualizing stress, and the viscosity of gases. He contributed theoretically to thermodynamics and statistical mechanics. Perhaps his greatest contribution was to unify electricity and magnetism in a set of four equations which showed that light was an electromagnetic wave, and laid the basis for all subsequent radio, electrical, and electronic technologies. His equations also provided the foundation for Einstein's work on relativity. In my opinion, no other discovery by any of Maxwell's contemporaries has made a greater difference to how we live now. The Nobel Laureate Richard Feynman put it even more strongly: 'From a long view of the history of mankind—seen from, say, ten thousand years from now—there can be little doubt that the most significant event of the 19th century will be judged as Maxwell's discovery of the laws of electrodynamics.'[1]

Maxwell had a strong Christian faith. Dying of cancer at the age of forty-eight, he was heard to recite from heart the whole of George Herbert's poem 'Aaron', 'Perfect and light in my deare breast, | My doctrine tun'd by Christ, (who is not dead, | But lives in me while I do rest)'. His science was integral to his faith. In his twenties he wrote,

[1] R. P. Feynman, R. B. Leighton, and M. Sands, *The Feynman Lectures on Physics* (Addison-Wesley, 1965), 1–11.

'Happy is the man who can recognize in the work of Today a connected portion of the work of life, and an embodiment of the work of Eternity.' Almost certainly at his instigation, a verse from the Psalms was carved into the entrance doors of the Cavendish Laboratory. The inscription is in Latin in an ornate gothic script.

When I joined the Cavendish for my PhD, the laboratory had moved from the original site in Free School Lane, which had become ever more cramped and inadequate, to new buildings in what is now the West Cambridge site for science and engineering. In my first year, I suggested to the Cavendish Professor, Sir Brian Pippard, that it would be splendid if Maxwell's inscription could be placed over the main entrance, and now that not so many physicists read Latin it should be in English. He agreed to put the proposal to the Policy Committee. Only years later did he tell me that he confidently expected them to veto it, and that it was greatly to his surprise that they enthusiastically agreed not only to the idea but also to the Coverdale translation.[2] Everyone who goes into the Cavendish Laboratory through the main entrance passes under this inscription, 'The works of the Lord are great, sought out of all them that have pleasure therein.' I love that as a motto for scientists. A photograph of the entrance forms the frontispiece of my PhD thesis. It says that what we are doing is finding out how God makes the world work, and it is enormously enjoyable.[3] In an incomplete draft found after Maxwell's death he wrote, 'I think Christians whose minds are scientific are bound to study science that their view of the glory of God may be as extensive as their being is capable of.'[4] It seems to me that, while any scientist can experience the intense joy of understanding something for the first time, those who know in some measure the creator whose works they are studying experience a hugely significant additional dimension of pleasure.

[2] A. B. Pippard, 'The Cavendish Laboratory', *Eur. J. Phys.* **8** (1987), 231–5.

[3] The Latin inscription, from the Vulgate, reads, 'Magna opera Domini exquisita in omnes voluntates ejus'. Thus, the pleasure is the Lord's. This translation owes much to the Greek Septuagint, which may well have been an accurate translation of a variant Hebrew version which is now lost. The Coverdale translation, and almost all subsequent English translations, follows the generally accepted Hebrew versions.

[4] L. Campbell, *The Life of James Clerk Maxwell: With a Selection from His Correspondence and Occasional Writings and a Sketch of His Contributions to Science*, ed. G. Garnett and M. Adams (Macmillan, 1882), 404–5.

My own journey of integrating my science within my Christian faith took on a new dimension through my friendship with the brilliant scholar and artist Roger Wagner, who has been described as 'Britain's best living religious painter'.[5] In 1997 Roger and I began a discussion about the nature of inquiry which culminated in a 500-page book.[6] In *The Penultimate Curiosity* we show how, where you have a culture, or a community, or occasionally an individual, with a curiosity about the ultimate questions of meaning and purpose, and how God can be known, and what it means to care, time and again that proves conducive to the kind of curiosity about the material world that has led to what we now call science. We use the metaphor of a slipstream. When birds fly in a 'V' formation, or Tour de France cyclists ride in a *peloton*, those behind don't have to work as hard as the one in front—they benefit from the slipstream. Studies of birds, such as the bald ibis, show that they find not only the best place in which to fly, but also the optimum phase for their wing beat, in order to maximize the energetic advantage.[7]

In a peloton, if the riders get too close the wheels can touch and cause a *chute*, which perhaps provides a further metaphor for the intellectual pileups which can occur when people try to use scientific methods to answer religious questions, or vice versa. But these are the exceptions, and Roger and I found far more cases where these different kinds of curiosity are closely entangled to mutual benefit.

From 2012 to 2015 the Director General of CERN, Rolf Heuer, convened a series of three workshops under the title 'The Big Bang and the Interfaces of Knowledge'. The subtitles were towards a common language, understanding of truth, and understanding of logic (an echo of A. J. Ayer's best-known title, see the opening of Chapter 5). In 2015, to mark sixty years of CERN, Rolf was invited to be guest editor of the Saturday Cultural supplement of *Le Temps*, the most widely read French language newspaper in Switzerland. The issue carried a full-page interview with me, largely covering what I had presented at the second conference. In his editorial comment on my interview Rolf wrote, 'Instead of opposing science and religion, you should rather ask if science and

---

[5] Charles Moore, *Spectator* (March 2014).

[6] Roger Wagner and Andrew Briggs, *The Penultimate Curiosity: How Science Swims in the Slipstream of Ultimate Questions* (Oxford University Press, 2016).

[7] S. J. Portugal *et al.*, 'Upwash Exploitation and Downwash Avoidance by Flap Phasing in Ibis Formation Flight', *Nature* **505** (2014), 399–402.

religion can respect and accept each other's point of view. In my view it is absolutely possible, and in my experience it is even enriching.'[8]

Can I hold together the rigorous scientific way of thinking that I learned from David Tabor and my other teachers, and the embodiment of the life of Jesus in the material world that I contemplated on Mount Tabor? Absolutely yes! In both science and religion there are hard problems which have not yielded easy solutions. More than one poll has revealed that the foundations of quantum mechanics remain hotly debated in the scientific community, with a divergence of views on some fundamental questions.[9] Scientists have not stopped using quantum mechanics until these questions are resolved; the theory is too robust, and too useful, for that. For rather longer there has been a diversity of views about foundational questions in religion, in such basic topics as divine action in a world well described by the sciences.[10] People do not stop exercising faith until such questions are resolved; religion is too robust, and too important, for that. Science not only elucidates the physical processes of the world, but also offers profound lessons in how we can live with unresolved questions without diminishing the quest for answers.

[8] R. D. Heuer, 'Au lieu d'opposer science et religion, il faudrait plutôt se demander si la science et la religion sont capables de se respecter et d'accepter le point de vue de l'autre. C'est à mes yeux tout à fait possible, et d'après mon expérience, c'est même enrichissant', *Le Temps*, Samedi Culturel (27 September 2014).

[9] M. Schlosshauer, J. Kofler, and A. Zeilinger, 'A Snapshot of Foundational Attitudes toward Quantum Mechanics', *Studies in History and Philosophy of Science B* **44** (2013), 222–30; M. Schlosshauer, J. Kofler, and A. Zeilinger, 'The Interpretation of Quantum Mechanics: From Disagreement to Consensus?', *Ann. Phys.* (Berlin) **525** (2013), A51–4; S. Sivasundaram, K. H. Nielsen, 'Surveying the Attitudes of Physicists Concerning Foundational Issues of Quantum Mechanics', arXiv:1612.00676.

[10] K. Ward, *Divine Action* (Flame, 1990); N. Guessoum, *Islam's Quantum Question: Reconciling Muslim Tradition and Modern Science* (I. B. Tauris, 2010).

7

The deeply subtle nature of physically existing things

Physics is the name we give to that part of scientific study which considers concepts such as physical entities (bodies) and forces, energy, mass, motion, space, and time. *Fundamental physics* is that part of physics which studies the basic structure of things, and how the simplest things move and interact with one another. This neighbours on metaphysics, because it includes thinking about the nature of time, and what can be said about the nature of physical reality.

One of the greatest transitions in human understanding of physics, arguably *the* greatest, took place in the early part of the twentieth century. This is the transition from what is now called 'classical' physics to quantum physics. In this chapter we explore the picture of the physical world that quantum physics provides. In the next chapter we shall draw out some lessons for our wider understanding. There are some useful lessons in the nature of knowledge and knowing, for example, and in the deep subtlety of physical existence. This can help us to manage the experience of living life in faith, without a full understanding of theological ideas such as the Trinity, for example. More importantly, quantum physics is an integral part of our role to study and celebrate God's creation, helping it to fulfil its true potential.

7.1 Classical and quantum physics

'Classical' physics brings together everyday notions of what physical objects are, with a simple but precise set of ideas about space and time and how objects move in response to forces. Examples of classical physics include the motion of an ordinary ball bouncing on the floor, or a planet orbiting the Sun, or things like bridges and car engines. The ideas needed to describe the motions in cases like these are largely as described by Isaac Newton, building on the work of people such as Philoponus, Grosseteste, Kepler and Galileo, Bacon, and a host of others. In the nineteenth century a further concept was introduced: a 'field', which is a continuous entity which can transmit force and energy. The electromagnetic field is the key concept in the study of electricity and magnetism.

Classical physics has, as its basic ingredients, readily imagined objects such as a ball or a rock or some liquid or a gas. In their large-scale properties, such things are much as we assume them to be from our immediate experience. That is, they each occupy some given position and have some definite motion at any given time, and their properties such as size and mass can be measured. They behave in a manner that may be hard to calculate but is easy to conceive of or picture imaginatively.

Quantum physics changes all this. The basic ingredient in quantum physics is not a ball or a rock or a liquid or a gas or any other readily imagined object, but a mathematical method whose correct physical interpretation is not universally agreed.

Let's introduce this by considering an example.

When we look at an ordinary wooden table, this is what we see. First, we see the surface, with its colour and texture and strength. Then we see the grain, and we begin to ponder the process which formed the wood, and we think about the plant cells, roughly a hundredth to a tenth of a millimetre across, of which it is built. Then we ponder the cellular machinery, no longer active in a dead piece of wood, but whose remains give the structures within the cells. Large organic molecules with many atoms of carbon and hydrogen are coiled and stretched into various shapes. All this can be imagined reasonably easily. It is a complicated network of parts, but the parts themselves appear to be simple things like balls or rods or sheets. But now when we go deeper, in each of our mind's eye we begin to see stranger things. Each molecule is made of atoms that are about one hundred thousand times smaller than a cell, and each atom is something like a ball, and yet something not at all like a ball, but more like a cloud or a trampoline. It has a hard nugget in the centre—the nucleus—and flying or swirling or vibrating around this nucleus is a cloud or a vibrance of electrons. We say 'cloud' because each electron does not orbit the nucleus like a planet orbiting the Sun, but rather enfolds the nucleus all at once, like a soufflé around a grain of sugar. We say 'vibrance' because this cloud is not vague but precisely tuned, exhibiting the most precise vibrations in nature, and causing a kind of superconduction of electric current around each nucleus.

The atom is both full and empty and the same time. It is full because if you tried to put another electron in, distributed around the nucleus in the same cloudy vibrating motion adopted by the electrons that are already there, then you will find it to be utterly impossible. There is absolutely no possibility of squeezing another electron in; it is forbidden by the very nature of what electrons are, and this is why solids are hard; it is the reason why our feet do not sink into the floor as we walk about. And yet, in another sense, the atom is almost empty, because it is very easy to pass other particles such as neutrons through the cloud of electrons, and other electrons can also pass through if they do so at high velocity. This does not contradict the previous statement because the spatial locations available to particles such as electrons depend on their state of motion. There can be more than two electrons occupying the same region of space, but only if they are moving in different ways. Such patterns of behaviour are utterly unlike what we ordinarily experience, and they cannot be accounted for in the terms offered by classical physics. The impossible-sounding features are owing to the fact that the basic components of all physical things combine and recombine like waves on the surface of an ocean of possibility.

All this, and more, is what we see in each of our mind's eye when we look at an ordinary physical object such as a cup or a spoon or a table top. The purpose of this illustration is to introduce the fact that when we get down to the electrons and the atomic nuclei and what goes on there, we find ourselves forced to contemplate a mode of existence which defies the categories of everyday language. It is a mode of existence that is precise in its mathematical details but eludes being expressed in any brief combination of words or physical concepts.

However, merely reiterating that quantum behaviour is rich and strange will not help the reader towards a more well-informed view of what this area of knowledge has to offer. In order to make progress toward such a view, we shall explain the way quantum physics 'works' in practice, in the sense of how its predictions are expressed and calculated. We shall do this by introducing two complementary mathematical methods, and comment on the picture of physical reality that emerges. The crucial ideas will be *waves of probability* and a special property called *entanglement*. It will also emerge that the concept of *information* and information processing is very useful in understanding quantum physics.

7.1.1 Quantum physics: approach 1 (the 'paths in spacetime' approach)

The first way to 'do' or calculate quantum physics works as follows.

First, we take a step back, as it were, and take a look at both space and time together. We note that all the various goings-on in the physical world can be seen as a patchwork or quilt of regions of space and time. The behaviour inside each region can be regarded as various particles coming in, interacting with one another, and then going on their way. Quantum physics describes this by furnishing a set of rules which allows us to calculate the probability for the particles to emerge at one place or another.

So far, so good.

But there is a subtlety. It is not straightforward to divide this quilt up into patches. If we cut the cloth too small, as it were, then the calculation goes wrong. This is owing to the fact that the probability of any given outcome can depend on several items coming together, sometimes reinforcing one another, sometimes cancelling one another out. This fact is called 'interference'. What this means in practice is that we must not assume, in any small patch in the quilt, that *either* one eventuality *or* another is the outcome; rather, we must allow that *both* may contribute to subsequent patches. For large enough patches, on the other hand, it is safe to say that some outcomes happen, and other outcomes do not—in agreement with our actual experience of life. For example, if you balance a pencil on its tip, and then allow it to fall, then quantum physics will allow for the possibility that it may fall in every possible direction, but in the event we observe it to fall in just one direction.

There remains a profound and persistent difficulty with this way of looking at quantum physics. It is that there is no simple or agreed way to determine how big is 'big enough' for these patches of space and time, to allow us to calculate the probabilities and say that we expect one outcome to happen, rather than all outcomes being carried forward together in combination, like many notes in a musical chord, or a ripple moving outwards in all directions at once. As a result, the boundary between one patch and another, or one set of events and another, has never been described in a way that everyone can agree upon; its nature remains open to various viewpoints.

Suppose, for example, that we have the event of an electron leaving the surface of a small narrow wire, going into the surrounding space, and ending up somewhere else. Think of an electron as something like a tiny tiny particle, but at the same time allow some vagueness into your picture; think of it also like an amorphous blob that can move out into space as an expanding cloud or vibration...you don't need to feel that you already know what an electron is; these pictures are just aids to the imagination. The idea is that, some time later, this electron is going to arrive somewhere else. According to quantum physics, there are a number of other places at which it could arrive, and we can calculate the probability of its arrival at each of them. But, and this is the mysterious part, we can't get a concrete picture of how this multiplicity of possible outcomes gets resolved into just the one outcome that occurs. It just does. That is the nature of things. That is what goes on in the world, every moment, everywhere. Millions upon millions of situations where, in each situation, a number of different events can take place, with probabilities that are given by the fixed patterns that God has guaranteed, and in each such situation, one of the possible outcomes happens, and no one knows any more about it than that.[1]

Of course, we do know more about it in the following sense: we have a good deal of knowledge about the fixed patterns that give the probabilities. These are described by the mathematics of quantum theory, with equations such as Schrödinger's equation, and beautifully rich descriptions of what we call 'quantum fields'. A quantum field is a physical thing, but one which stretches everywhere and forms part of the fabric of space and time. It can oscillate and thus contain various amounts of energy, as well as other properties such as electric charge and momentum. These properties can move around in the field like waves on the surface of the ocean. The electron that we mentioned above is regarded as a way of talking about the behaviour of one of these quantum fields. Or, if you like, you can look at it the other way around and say the fields are ways of talking about the behaviours of electrons (and other entities such as quarks and neutrinos). A large number of further fundamental particles also have associated fields and the combined structure of these fields is both rich and beautiful, like a tapestry shimmering with a variety of patterns all interleaved together.

---

[1] We postpone to the next chapter a presentation of existing attempts to resolve this. Our claim that 'no one knows' is legitimate because no existing proposal commands a consensus of well-informed opinion.

7.1.2 Quantum physics: approach 2 (the 'state in Hilbert space' approach)

Next, we present another way to think about all this. This second way to consider quantum physics is the way it is often done in practice in science laboratories, perhaps because people find it conceptually easier. In this approach, the first step is to focus one's mind on something called the 'quantum state', which means 'state of affairs' or possibly 'state of knowledge' or possibly just 'a mathematical abstraction useful for calculating what physical things will do'. There is no scientific consensus on which of these descriptions best captures what we mean by a 'quantum state'. These positions, and others, are held and argued-for at length by physicists and philosophers in their efforts to think it through carefully; it may be that several such positions have elements of truth and none exhausts the truth.[2]

The best way to think of a quantum state in the first instance is possibly just to assign it a mathematical symbol, and supply rules about how to manipulate the symbol. The standard symbol is

$$|A\rangle.$$

That is, a vertical line followed by a letter (often Greek, but it doesn't matter; here we use a Roman letter), followed by an angled bracket. There are then rules which describe mathematical operations on this symbol; these are the rules that quantum physics provides. They tell us that this symbol stands for something that behaves much like an arrow that can rotate, but these rotations take place in an abstract mathematical space that has multiple dimensions (there can even be an infinite number of dimensions!). The rate of rotation is given by a bunch of equations which depend on the circumstances, and the final ingredient is where the mystery comes in again. There is a rule for how to get the probability that, given that the state is now $|A\rangle$, then all future observations will be consistent with it being, in fact, $|B\rangle$. Here, $|B\rangle$ is not necessarily the same as $|A\rangle$, though it will be pointing in a somewhat similar direction. You see, whenever the development from the past says the arrow is now pointing in the A direction, there is a chance, a well-defined probability that we know how to calculate, that all developments in the future will nevertheless be just as if the arrow is now pointing in the B direction! So which is it? A or B? The past says A; the future says B. Nobody knows how to resolve this. We just say that the

---

[2] G. A. D. Briggs, J. N. Butterfield, and A. Zeilinger, 'The Oxford Questions on the Foundations of Quantum Physics', *Proc. R. Soc. A* **469** (2013), 20130299.

jump from A to B occurs randomly and we know how to calculate the probability that it will occur in any given case.

Here is an example. Suppose an electron leaves a small wire, as in the example we briefly gave earlier. We can arrange an experiment, using a magnet to provide a force on the electron, such that the electron accelerates either up or down, in the vertical direction. The arrow associated with the quantum state in this example would point up when the electron accelerates upwards, and point down when the electron accelerates downwards. But it often happens that the arrow (the mathematical abstraction) points sideways. What does that mean? Does the electron move sideways? No, because in this experiment the motion is constrained to be in the vertical direction. When the mathematical arrow points sideways, it means the electron in some sense goes both up and down. To be precise, if there are detectors in each path (one above, one below), then only one detector receives an electron, and the probabilities of which detector it is are equal (50:50 as we often say).

This observation might not puzzle us too much; one could suggest, for example, that what really happened was that the electron went one way or the other, not both, and we have the probabilities for each outcome. However, further experimental evidence contradicts this simple interpretation.

Quantum physics was developed initially in response to various puzzling experimental observations, such as the stability of atoms in states of discrete energy. We have focused so far on the fact that it gives a physical picture which remains puzzling. But this has not prevented it from being extremely insightful and useful. Its essential ideas underlie or make sense of a vast range of experiments, and open the way to understand chemistry, materials science, and a large number of useful devices such as the transistors that fill computer chips, and lasers, medical scanners, and so on.

7.2 Entanglement

Next, we will try to convey an important aspect of the overall picture of physical things and phenomena that quantum mechanics provides. This is an attempt to convey what things are like at their most fundamental level, the level of their very existence in space and time. Such issues may have a bearing on questions of human identity and the mind–body problem. They are certainly relevant to gaining a correct understanding of the way the world is.

In the past it was possible to imagine that the world consisted of lots of little ball-like particles jostling and careening in space, and thus, in their weaving motions, making up all the material of the world: the cup, the spoon, the blade of grass, the wings of the humming-bird, the brain of a sleeping child. This picture has elements of truth but is far from fully correct. According to the vast array of experimental observations, calculations, and technology that we call quantum physics, this picture doesn't work. The stuff of the world is not like lots of little bits and pieces that join up with one another like a myriad myriad of Lego bricks. The stuff of the world is more like overlapping waves, and it is subtle, forever refusing to yield up an easy summary of its own nature when we try to put our experimental finger on it.

It can hardly be overemphasized how great is the change in our understanding of physical reality that takes place in the transition from classical to quantum. It is the change from the Lego-brick picture to the dance-of-the-probability-waves picture.

A very important and profound feature of this change is that it introduces into fundamental physics a limit to the validity of *reductionism*. Reductionism is the philosophical term for a simple and very widely used idea in science, namely that the whole is the sum of the parts. This is the idea that a composite thing can be understood by regarding it as a collection of parts, each part having its own self-contained nature. In order to understand the working of a mechanical clock, for example, one would examine the various gears and springs, and bring in the fact that each part pushes or pulls on its neighbouring parts, and that is all there is to it. There is no further ingredient in addition to these separable parts and their interactions. Until the advent of quantum physics, it was widely assumed that the natural world is like this, at the level of its basic bits and pieces. It now appears that that is not quite the whole story.

One should not exaggerate this aspect of quantum physics: reductionism remains a very important insight into most of physics, and it is only partially qualified by some aspects of quantum physics. It is not removed altogether. But the fact that it is qualified at all is a startling insight into the nature of the physical world.

The aspect of quantum physics where this qualification is seen is called *entanglement*. In order to explain what entanglement is, we will present an extended example. We already explained that quantum states are represented by abstract mathematical symbols which can point like arrows. When two physical things are considered together, such as two electrons, then the arrow that represents the quantum state of both of them together has more directions in which it can point. To be precise, it has more *dimensions*, like the change from a flat circle to a solid sphere. This represents a richer set of quantum states.

How should we imagine this richer set of states? For many purposes, it is enough to say that we have two separate electrons, each with its own properties, and we are studying how they behave. But it turns out that for many of the quantum states, this way of describing them cannot work. We cannot simply say that we have 'two electrons'; we can only say we have *one pair*. It is a bit like the distinction between the topology of two separate rings and two interpenetrating rings. Or it is like saying that when two people get married, there is one married couple. It is no longer appropriate to describe them as altogether separate with their own agendas. Similarly, for a pair of entangled electrons, it is no longer appropriate to describe them as altogether separate with their own properties. This leads to phenomena such as electric current flowing without resistance (superconductivity) and various other materials properties, and it is a central aspect of quantum computing which we will describe in Section 7.4.

To understand what entanglement is, let's take two very small magnets (in an actual experiment, each magnet might be a single atom of an element such as calcium). We place these magnets next to each other in a chamber, but not too close, and arrange for them to point downwards. This is done by a sequence of laser pulses; the details are not important. We have good evidence that the sequence works and the magnets do indeed point in the direction we think, because we can then measure them and find out. We thus get confidence that this *state preparation procedure* is reliable. Next, having prepared both magnets to point down, we turn one of them over, so that it points up (see Figure 7.1). This is done by allowing a short pulse of laser radiation to illuminate the chosen

Figure 7.1 Two magnetic atoms side by side, one with its north pole up, one with its north pole down.

calcium atom. The electromagnetic field in the pulse connects to the magnetic property of the atom, causing it to rotate, and if the timing is done exactly right, then the rotation is through 180°. Then, we measure the outcome, and we try this experiment very many times (in fact, millions of times) to be sure that we know what we are doing. This is possible because these experiments can be repeated very quickly in a modern automated laboratory. Indeed, experiments of this type are regularly done in the laboratories of two of the authors (AS, AB). In this way we can prepare the magnets pointing up or down as we like.

Next, we explore some further possibilities. We can examine cases like the electron experiment we've just described: cases where a magnet is pointing both up and down (roughly speaking). In these cases, at the measurement stage we observe 50:50 results. Half the time we find the magnet pointing 'up' and half the time we find it 'down'.

Finally, we do the following experiment. We prepare the magnets in the special joint state called an *entangled* state. Our aim is to study what sort of state this is, by manipulating and measuring it. We can measure whether the magnets are up or down, and we can flip them over by laser pulses. The experiments are repeated many times and everything is checked very thoroughly. The main result is easy to describe:

If you have the magnets in the special entangled state, then turning either one of them over changes the state, but turning both of them over does nothing whatsoever to the joint system.

From both an experimental and theoretical point of view, flipping both magnets does nothing at all to the quantum state in this case.

That observation might not seem too remarkable, but now let's think. For brevity, let's agree that E represents the state the magnets are first prepared in. Next, suppose we flip (that is, rotate through 180°) the magnet on the left. The state is now 'E with a left-magnet-flip'. Let's call that E-L. Now suppose you want to return the system state to E. How would you do that? Easy: just flip the left-hand magnet back again. That works, but what is very interesting is that you have a choice. Because instead you could flip the right-hand magnet, thus getting the state 'E with both magnets flipped', *and this is the very state that is found to be identical to* E.

Now we have arrived at the heart of the matter. In the case of an entangled pair of magnets, in order to make the *given* change to the total system (moving from E-L to E) you can apply an operation *either* to the left magnet *or* to the right magnet. It takes some imagination to see

Figure 7.2 Two magnetic atoms side by side, with their magnetic property in an entangled state. The diagram is an attempt to visualize the situation, but no picture can do this very well. The magnetism (indicated by the doubled-ended arrows) is located at the atoms, but it is impossible to assign it to the atoms separately

how remarkable this is. It is like saying that if a stool in the kitchen has been turned upside down, then in order to correct the situation I could flip over some other stool in the dining room. Or it is like saying that if the front tyre on my bike has a puncture, then in order to mend the bike (returning it to a pristine condition), instead of mending that tyre, I could instead puncture the other one! These comparisons show that the magnets in the entangled state are not behaving like two ordinary objects such as stools or bicycle tyres. They are behaving another way.

The way we can make sense of this is to say that in the case of the entangled state, the two magnets are no longer separate entities each with their own properties. Rather, their directional (up/down) property has formed a single whole, and operations applied to either are operations applied to the whole (see Figure 7.2). This statement in words corresponds directly to the mathematical description of the entangled state. The state cannot be expressed as a product of two separate parts. This mathematical fact is reminiscent of the distinction between a composite number and a prime number.[3] It can be shown mathematically that *the entangled state cannot be described using any description that tries to model the system as two individual things.* And furthermore, this remains true

---

[3] A composite number is a positive integer that can be formed by multiplying together two smaller positive integers. A prime number is a positive integer that has no divisors other than one and itself.

even when the magnets themselves are in identifiably different places. In the description above we said simply 'one on the left, one on the right', but this could equally be 'one in London, one in New York' or even 'one on Earth, one on the planet Mars'.

A pair of entangled atoms, or entangled molecules, or entangled parts of proteins, or larger things, is in a state that cannot be described, in full, using our everyday notions of what things are—the notions that we use in our everyday thinking and imagining, and which are used in classical physics.

These sorts of states are now the subject of a large amount of experimental research effort, to check that they do indeed behave as quantum theory says they behave, and also with a view to seeing whether they play significant roles in other natural phenomena, such as in the structure and properties of solid materials, or in the chemical processes in living cells. It is not self-evident that they will, because the phenomenon of entanglement is very fragile: very small disturbances are sufficient to disturb the delicate balance that an entangled state requires. However, some theoretical studies have suggested ways in which a protein, for example, could exhibit long-lived entanglement when the biological processes in a living cell prevent it from reaching thermal equilibrium. This is an exciting ongoing research area.

We shall discuss some of the implications of entanglement for our understanding of other things after we have presented some more components of quantum physics.

7.3 Quantum information

In their eagerness to understand, physicists have pondered long and hard about what a quantum state *is*. However, it is interesting to note that it is often better to ask not what a quantum state *is*, but what we can *do* with it.

For about sixty years, between 1930 and 1990, quantum entanglement was an intriguing phenomenon, but not one for which any significant uses were known. This began to change when, building on seminal work of John Bell in 1964,[4] physicists began to think about quantum states more and more in terms of their information content. Here we have in mind the technical definition of 'information' that is used in computer science,

[4] J. S. Bell, 'On the Einstein Podolsky Rosen Paradox', *Physics* 1:3 (1964): 195–200.

where 'information' is a measure of how much digital memory would be needed to store, or faithfully transmit, a given document or picture, or other such information-bearing entity. In quantum physics, we use the same definition, but now the memory is made of quantum systems. These quantum systems can explore all the range of quantum states available to them, including entangled states, and thus 'quantum bits' have a richer behaviour than the more simple binary bits used in ordinary computers.

Quantum memory capacity is measured in units called 'qubits' (pronounced 'queue-bits'). You can think of one qubit approximately as one electron, or one photon, or one atom. Actually, a single atom can often store more than one qubit, but it is useful to keep in mind that qubits always mean quantum states of physical things such as atoms, not just mathematical abstractions.

A central consideration in information storage, transmission, and processing is how to protect the information from copying or transmission errors, or from degrading over time. One way is to record the information in multiple copies. Then a mistake can be detected by comparing copies and using a majority vote. This has been the main way in which treasured documents such as the Bible have been preserved, for example. However, in the case of quantum states, there is a difficulty. It is not possible to make perfect copies, because one cannot determine what quantum state one has without first measuring or observing it in some way, and *the observation process itself disturbs the qubit being observed*. For this reason, faithful copying of a general quantum state is rendered impossible, and therefore the standard method of preserving information is not available. As a result, the whole business of protecting data from errors becomes an important and non-trivial issue in the physics of quantum information. Because the disturbance caused by measurement is a fundamental and unavoidable fact about the laws of quantum physics, it led physicists to think that the laws of nature are such that large entangled states (involving more than a few qubits) must always be inherently fragile, and not capable of being made more robust.

It turns out that this is not so. There are ways to protect quantum information from errors, by a process that does not involve copying, but which works by enlarging the data storage system in a more subtle way. The crucial insight is that of a *quantum error correcting code*. This is a name for a highly patterned set of quantum states with certain specific properties.

The existence of these states or 'codes' was discovered independently by Peter Shor and Roger Calderbank working in America and by one of us (AS), working in the UK.[5]

A 'quantum error correcting code' is a set of 'quantum codewords', and each quantum codeword is a very specific highly entangled state involving many qubits. These states are also highly fragile, but one can cleverly construct them so that a very ingenious type of observation process is possible. This observation process, called error correction, is a measurement that reveals *not what state you have, but what error has occurred.*

How is this possible? We will describe the flavour of the idea by means of an analogy. The whole strategy can be compared to using musical chords instead of single notes to record a given tune. If we agree at the outset that only major chords will be used, then it becomes possible to detect whether there is a note wrong, whenever the wrong note results in a change from major to minor. What happens in quantum error correction is that the qubits are measured in such a way that the measuring apparatus determines whether the chord is major or minor, without determining which chord it is. This is what allows the correction process to detect and correct the errors without measuring and thus disturbing the music itself.

We have presented the idea of a quantum error correcting code partly for its innate interest, and partly in order to illustrate the move from the question 'What is it?' to the question 'What happens to it?' in quantum physics. This change of question moves the focus of interest from *things* to *operations.*

It turned out that this same change of emphasis from things to operations could also be applied to the codes themselves, and this yielded much new insight.

The discovery of quantum error correcting codes involved the ability to 'see' in one's mind's eye the notion that we briefly illustrated by analogy to musical chords. In the case of the states called quantum codewords this involves picturing a set of interleaving crystalline structures in a high-dimensional abstract mathematical space. Not something

[5] P. W. Shor, 'Scheme for Reducing Decoherence in Quantum computer Memory', *Phys. Rev. A* **52** (1995), R2493-6; A. M. Steane, 'Error Correcting Codes in Quantum Theory', *Phys. Rev. Lett.* **77** (1996), 793–7; A. R. Calderbank and P. W. Shor, 'Good Quantum Error—Correcting Codes Exist', *Phys. Rev. A* **54** (1996), 1098–105; A. M. Steane, 'Multiple Particle Interference and Quantum Error Correction', *Proc. R. Soc. Lond. A* **452** (1996), 2551–77.

anyone can readily visualize! Because this is hard to imagine, the code-words themselves are hard to work with. However, a young researcher at the boundary of physics and mathematics, Daniel Gottesmann, suggested that we should think of quantum codewords another way. Instead of trying to write them down or picture them, we define them in terms of operations applied. One first proposes a group of operations, such as 'rotate the first and third qubits, and flip the seventh qubit'. Then a codeword can be defined as 'that quantum state which does not change under this group of operations'. Whole sets of codewords can be defined in one go like this. The emphasis now is on constructing suit-able sets of operations.

This approach was not only more straightforward, but also yielded several new insights, and it led immediately to finding new types of code. Hence the change of focus from 'What is it?' to 'What happens to it?' proved again to be very fruitful. The move from thinking about states to thinking about operations is like a change of focus from form to function. A similar change can be fruitful in understanding some religious ideas; we shall return to this point.

7.4 Quantum computing

As soon as the idea of thinking about quantum states in terms of their information content proved to be fruitful, it was natural to ask what sort of information processing or computing might be possible with qubits. What would a computer based on quantum devices be like, and what could it do? Would it simply be small and fast, but operate on the same principle as current computers, or would quantum physics, espe-cially entanglement, make possible a qualitatively different form of information processing?

Two ways of conceiving the design and function of a quantum com-puter were described by Richard Feynman and by David Deutsch in the 1980s. Both involved a set of simple entities such as single atoms held in an array. Feynman considered what could happen if the interaction between each atom and its neighbour could be controlled in some way, so that one could 'order up' any desired interaction that was allowed in principle by quantum physics. Deutsch simplified the problem in a powerful way, by thinking of using a finite set of operations called 'quantum logic gates', and showing that this would be sufficient. A 'logic gate' is a fixed process applied repeatedly to chosen groups of atoms. This specified the computer sufficiently fully to allow its capabilities to be explored.

The quantum computer was thus first described as a design concept or a blueprint. The 'input' to the computer is the initial state preparation, the 'program' is a sequence of logic gates, and the 'output' is a measurement of the final state.

Once this way of thinking about quantum computing had been discovered, it opened the way for further theoretical work which addressed the question, would such a machine be good at computing? Could it outperform other types of computer? It turns out that the answer is *yes*.

Assessing the speed or efficiency of computing devices is itself a nontrivial task, and we will not describe it in full. However, the main idea is very simple: when considering whether a computer can solve a given task, one must take into consideration how long it would take. It is all very well saying that my machine 'can' calculate a given problem, but what if, in fact, it would take billions of years to complete the calculation? In that case then the computer cannot handle the calculation in any practical sense. The difference between quantum computers and traditional computers is that there exist some special (but significant) tasks where the quantum computer performs so much more quickly that it makes the difference between practical possibility and impossibility.

For interested readers, we will now flesh this idea out a little, but one could skip to the last paragraph of this section without missing anything essential.

A good way to illustrate what is involved in assessing the speed of computers is to consider the task of looking up words in a dictionary. In a traditional dictionary printed on pages in a book, we can find a given word reasonably quickly because the words are arranged in alphabetical order. Imagine a dictionary in which the words were given in random order. It would be almost impossible to use! Every time you wanted to look something up, you would have to search laboriously through all the entries until you spotted the one you needed. It is not that it can't be done, but it would take too long.

The same reasoning applies to mathematical tasks. Some are quick, like finding an item in an ordered list; some are slow, like finding an item in an unsorted list.

Suppose you have a long sequence of numbers that is random for some finite length, and then after that repeats that same random sequence over and over. Such a sequence is like a burst of noise that keeps repeating over and over again. One would like to discover where the repeat occurs. A standard computer could find the repeat by going through the sequence one number at a time. This is slow. A quantum computer can tackle this problem in a clever way which involves an entangled state where one part of the memory stores the sequence, and another part records where in the sequence each item occurs. By manipulating this quantum memory in the right way, the repeat period can be obtained without having to look at all the items one by one. The discovery of the 'quantum algorithm' that achieves this is a substantial development in our mastery of quantum behaviour and hence of physics. It is owing to a sequence of insights, with notable contributions from Richard Jozsa, Daniel Simon, and Don Coppersmith, culminating in a famous paper by Peter Shor.[6]

In the one-by-one method that a classical computer must use, if the length of the sequence doubles, then so does the calculation time required to find the repeat. In the case of a quantum computer the situation is very different. Now if the length of the sequence doubles, the solution only takes a little longer. It is like replacing a given dictionary by one with twice as many pages: this doesn't change the process of looking something up very much. To be precise, for the quantum computer the time increases in proportion to the logarithm of the length of the repeating sequence under study. This is a very great improvement over traditional or classical computing.

To see how great a difference this makes, consider the following example.

Let us suppose, for the sake of argument, that an ordinary computer can generate and examine some particular sequence at the rate of one thousand integers per second, whereas a quantum computer can operate only at a much slower basic rate, owing to all its associated control apparatuses, such that it takes ten minutes to construct and examine a sequence of one thousand integers. In this case the quantum computer offers no practical benefit when the sequence is one thousand integers long. So let's consider a longer sequence: say, one million numbers. Now the ordinary computer takes one thousand seconds, which is about seventeen minutes, and the quantum computer takes twenty

⁶ P. W. Shor, 'Polynomial-Time Algorithms for Prime Factorisation and Discrete Aogarithms on a Quantum Computer', in *Proceedings of the. 35th Annual Symposium on Foundations of Computer Science* (IEEE Computer Society Press, 1994), 124–34.

minutes. The ordinary computer wins again, but only just. When we go to sequences longer still, the quantum computer will win.

At sequences of one billion numbers, the quantum computer takes thirty minutes, while the ordinary computer takes one million seconds, which is eleven and a half days. This reveals the significance of the fact that the number of steps required by the quantum algorithm increases so much more slowly with the size of the problem. And what is really interesting is that when we go further still, and consider longer problems, the quantum computer is able to complete the task in a reasonable time, whereas the traditional computer is simply too slow to be practical. For example, in the repeating sequence problem, if the sequence is long enough that the quantum computer takes forty-five minutes, how long would an ordinary computer take? Answer: one thousand years!

The repeating sequence problem is just one of a number of useful calculations where quantum computers can solve the task while ordinary computers cannot. In fact, this particular problem is more important than one might think: it can be used to find the prime factors of large integers, and this, in turn, allows one to break various encryption protocols widely used for secure communication throughout the world. However, the main point is that there are various more constructive tasks, such as aiding scientific research and solving pattern recognition problems, that quantum computers would also be very helpful with. For these tasks the power of a quantum computer increases exponentially with its size.

There are three main lessons we can draw from this. First, quantum computers would be very useful computing devices, and this is why efforts to make them are under way in many research laboratories. Second, we have here a practical use for quantum entanglement. Third, there is something going on in quantum physics that is different from ordinary digital information processing.

7.5 Brain function

It is very natural to ask whether quantum physics, and quantum entanglement and computing in particular, might have something to say about the functioning of the human brain. The question as to what role quantum processes play in life was posed by Erwin Schrödinger in some lectures at Trinity College Dublin.[7] Sir Roger Penrose has speculated

---

[7] Erwin Schrödinger, *What is Life?* (Macmillan, 1944).

about quantum processes in the brain.[8] At the time of writing, the jury is out on this question, and likely to remain so for some years.[9]

Anyone with a reasonable knowledge of physics and biology can form a hunch about this, and some estimates have been made. One way to assess it is to estimate whether an entangled state can survive long enough to impact significantly on a process in a neuron or between neurons. Most such estimates imply that phenomena related to entanglement are too short lived in liquid- or solid-state processes at room temperature, and therefore entanglement has no relevance to brain function. This is what the majority of neuroscientists and physicists probably think at present (we don't know of any definite survey of opinion on this; we are only reporting our experience of chatting with fellow workers in this area.) However, the idea that there may be a role for quantum entanglement in brain function is a perfectly respectable scientific idea that is the subject of ongoing study. This should not be regarded as an excuse to indulge in idle pseudoscientific speculation on this question; rather, it is a motivation for well-designed experimental work in the lab, and for well-thought-through theoretical suggestions of what to look for. An example is a theoretical study which suggests that when the right kind of driven oscillation is present, some forms of entanglement can survive for long periods under conditions where thermal noise would otherwise cause the entanglement to disappear very rapidly.[10] This is enough to show that one can expect to find some role of entanglement in a biological context such as in electron transport in proteins in a living cell, but it is an open question whether this happens in such a way as to significantly influence the functioning of a cell.

---

[8] R. Penrose, *The Emperor's New Mind* (Oxford University Press, 1989); *Shadows of the Mind* (Oxford University Press, 1994).

[9] Many of the developments and questions were brought together in D. Abbott, P. C. W. Davies, A. K. Pati (eds), *Quantum Aspects of Life* (Imperial College Press, 2008).

[10] G. G. Guerreschi, J. Cai, S. Popescu, and H. J. Briegel, 'Persistent Dynamic Entanglement from Classical Motion: How Bio-molecular Machines Can Generate Nontrivial Quantum States', *New Journal of Physics* 14 (2012), 053043.

8

Issues arising from quantum physics

We now turn to some issues that arise at the boundary of quantum physics and philosophy and theology. First, we present the fact that the basic aspects of the nature of the physical world are not agreed upon by the scientific community. Then we present some issues relating to personal identity, whether for human beings or other conscious or responsible agents. Then we comment briefly on whether quantum physics presents new possibilities for how God acts in the world. Finally, we present some positive lessons for ordinary followers on the way of Christ.

8.1 Interpreting the picture of physical reality

Let us take a step back and try to consider what is the picture of the physical world that emerges from quantum theory. This is not easy to do. Throughout the past century since its mathematical principles were first worked out, there has been debate about the physical interpretation of quantum physics. What does quantum physics show us about the nature of physical reality?

So far we have pointed out that, whatever interpretation one adopts, one is contemplating a universe quite different, in its fundamental nature, from the picture given by classical physics. This difference is shown most directly in the phenomenon of entanglement.

Among the problems of interpretation lies the question of the evolution from past to future, and what is called 'the collapse of the wave function'. In the previous chapter we pointed out that experimental evidence leading up to any given moment may indicate that the quantum system evolves to some state $|A\rangle$, whereas observation of the outcome is consistent with the state being $|B\rangle$. State $|A\rangle$ might be described as 'electron travels both up and down' and state $|B\rangle$ might be described as 'electron travels up, and only up'. The change from $|A\rangle \rightarrow |B\rangle$ is called the 'collapse of the wave function'. The problem is that this collapse does not have the character of a physical process following the equation of motion provided by the theory; rather, either it is owing to some new physics or it is just a human convention, that is, a way of speaking about a situation which perhaps could be better described in other terms.

There are, broadly speaking, three ways to react to this situation, which result in three broad classes of interpretations of quantum physics.

Interpretation 1. We accept that the mathematical symbols only connect indirectly to the physical situation. The mathematical apparatus does not tell us directly the physical state of the world; it tells us how to calculate the relative probabilities of the various possible events. The collapse of the wave function is not a physical process, but a convenient way to help our imagination when we think about the development from past to future. Among all the possible final states available from any given starting point, just one is adopted. (The interpretation advocated by Niels Bohr and the Copenhagen School lies in this class.)

Interpretation 2. The mathematical quantum state corresponds directly to the physical state, and the collapse of the wave function is again only an appearance. What in fact happens is that the universe has one vast quantum state that evolves smoothly, and the resulting structure of physical reality is like an exponentially branching multiplicity of universes which we inhabit without being directly aware of it. (Hugh Everett was an early advocate of this view).

Interpretation 3. The collapse of the wave function is a real process that happens by virtue of further physical effects as yet not well understood. (Various people have suggested types of process that might be at work.)

Interpretation 1 here is a perfectly good way forward, but it suffers from the weakness of being somewhat vague, or else, in some versions, mathematically inelegant. Interpretation 2 is called a 'relative state' or 'many-worlds' interpretation and in some respects is easier to define and therefore study. Interpretation 3 has also been much studied, in terms of suggestions as to what the mechanism of the collapse process might be, and experimental tests have begun to probe this.

The problem with adjudicating between these interpretations is that it is very hard to say which one best respects the principle of Occam's razor, which is the principle that among accurate explanations, simpler ones are to be preferred. Which of the interpretations, if any, are accurate and which is simplest? In the first interpretation, there is an awkward vagueness surrounding the sense in which the mathematical apparatus captures or expresses the physical situation. This does not affect the use of the theory in practice, but it might be said to be an unwanted complexity. In a many-worlds interpretation, the basic mathematical apparatus is simple, but that which is said to be 'real' or 'have physical existence' is very much more complex, because it includes structure that is like copies of universes repeatedly doubling and doubling again at a very fast rate. All of planet Earth, including all its biosphere, and the Sun and the galaxy, and so on, is, according to this, currently existing in a vast and ever-doubling collection of copies that are unobservable to us, but observed by copies of us living in them. The evidence that these copies exist is very indirect, however. The evidence consists purely and completely in the fact that if multiple unobservable copies of the universe do exist then the mathematics of quantum theory is, for some tastes, more elegant.

We have described this interpretation problem in order to alert the reader to it, but since the interpretation problem in quantum physics has been going on for a long time and does not show signs of an emerging consensus,[1] we feel that the best attitude now is to remain agnostic on the three interpretations, and continue to seek new data. However, we should add the following note of caution. The many-worlds interpretation raises many deep questions of personal identity and moral action which have hardly begun to be thought out. This raises further difficulties, alongside the one of Occam's razor, that will affect the interpretation of the theory. Consider the following scenario. A couple are on the beach, watching their son swimming in the sea. The son gets into difficulty. The parents argue about whether he needs help, and if so which of them is the stronger swimmer to go to the rescue. As a result of their procrastination they are too late, and their son drowns. It would be of little consolation, later in their lives, to ruminate that there may be another universe in which one of them rescued their son and he is now alive and well.

---

[1] M. Schlosshauera, J. Koflerb, A. Zeilinger, 'A Snapshot of Foundational Attitudes toward Quantum Mechanics', *Studies in History and Philosophy of Science Part B: Studies in History and Philosophy of Modern Physics* 44 (2013), 222–30.

When such issues of personal identity and responsibility are explored, as they will be in the absence of further empirical data, one should keep in mind that there are currently no experimental data which support one of the three classes of interpretation of quantum theory above the others, and furthermore there is no way to construct an experiment which could adjudicate between interpretation 1 and interpretation 2. In short, the idea that all possible developments of the universe coexist and multiply in a vast multiverse is a conjecture for which there is no experimental evidence that is not also consistent with the existence of just one universe: the one we live in, observe, and discuss.

8.2 Implications of quantum physics for issues of personal identity

Now we turn to issues of personal identity. Or to put it more colloquially, *can we learn anything about the human soul from quantum physics?* The answer to this question is, we would like to argue, 'yes and no'. One should be cautious of leaping too quickly to theological arguments from observations about the nature of the physical world.[2] However, the physical world is no longer thought to be the sort of deterministic clockwork mechanism that it was imagined to be in the past by some leading voices in science and philosophy, and we can dismiss any arguments that were based on that supposition. For example, it is not possible to argue, on the basis of fundamental physics, that human beings are automata with no responsibility for their actions, because our understanding of the human brain is nowhere near complete enough to venture any such conclusion. However, it is also not possible, from fundamental physics alone, to show that humans are responsible agents. The question is, from a scientific point of view, open. What one may claim, with good intellectual credentials, is that the notions of human free will and responsibility are not ruled out by what we know of quantum physics, as long as we add the following important remark: *we (the human race) have really very little idea how this works in practice, and maybe it involves aspects of the basic physics of the world that are as yet undiscovered.*

---

[2] L. Campbell, *The Life of James Clerk Maxwell: With a Selection from His Correspondence and Occasional Writings and a Sketch of His Contributions to Science*, ed. G. Garnett and M. Adams (Macmillan, 1882), 404–5.

It is a mistake to suppose that mere indeterminacy in physics is itself sufficient to resolve the philosophical difficulties surrounding human free will. But the unknowns that remain in both philosophy and neuroscience, and the subtlety of physical reality that quantum physics shows us, together make it correct to say that the notion of *human responsibility* is not ruled out by our scientific knowledge of the world.[3]

8.2.1 The soul

What is meant by the soul, in standard contemporary Christian theology, is that which makes a person who he or she is, in sum total. It is a concept which gathers together all that goes to make up a human being. When we read an obituary, or hear a funeral address, we are typically hearing an attempt by one person to say who another person was in their own nature and in the contribution they made; this is what 'soul' is about.

In its correct usage, the word soul is not a further attribute that humans have in addition to their physical attributes. Rather it affirms the precious nature of each person, owing to the fact that their choices are meaningful, not just mindless responses to stimuli. To say of a human that they are a soul is similar to saying of a sequence of sounds that it is music. Music is not a further physical attribute in addition to the sounds; it is a statement of what the sounds amount to. Similarly, your soul and mine are simply who we really are. To 'save' a soul is to act to prevent a soul from becoming soulless; that is, to prevent a slide into self-centredness.

It is an absolutely central pillar of Christian thought that souls are precious. There are three ways in which quantum physics may bear on this.

First, according to the physics of quantum states, it is not possible to build a universal cloning machine. That is, the kind of copying process that we are familiar with in printing, photocopying, and electronic file-sharing is not possible when it comes to the complete details of quantum states. If I have a physical system such as a water molecule in some state $|A\rangle$, then you may want to take another water molecule and prepare it at another location in that same state $|A\rangle$. If you did this without disturbing my copy, then it amounts to a copying or cloning process. However, according to the rules of quantum physics, it cannot be done.

[3] H. J. Briegel, 'On Creative Machines and the Physical Origins of Freedom', *Scientific Reports* 2 (2012), 522.

This is called the *no-cloning theorem*. The pertinent point is that if you did not already know what state my molecule was in, then in order to prepare a copy you would have to observe my molecule in some way, but this observation disturbs the state, so that you cannot be sure of getting, as the end result, the very same state that my molecule started in.

At the time of writing, we know that the no-cloning theorem is true of the details of the quantum states of things such as molecules, but it is not known whether this bears on human identity. But it might. It is not known whether brain function depends on the details of the quantum states of some of the molecules and larger structures in our brains. If it does, then the type of copying process invoked by science-fiction writers, when they imagine that people could one day save 'backup' copies of themselves, may be physically impossible. We already know that each person is unique. It may be also that each person is intensely fragile.

The no-cloning theorem does not prevent a quantum state being passed on from one physical embodiment to another, as long as the first embodiment loses the state when the other one gains it. One way this can happen is through a process known as quantum teleportation.[4] This would be like the human faxing thought experiment described in Chapter 5, except that if entanglement is involved, then not pressing the self-destruct button is no longer an option. Thus, the notion of backup copies might prove to be incoherent when it comes to persons, but there is nothing here to prevent the Christian idea of Resurrection from having intellectual coherence.

A second aspect of quantum physics that bears on the soul is the limit to reductionism that we presented in the context of quantum entanglement. Once again, the status of our scientific knowledge is such that we do not know whether quantum entanglement is significant to brain function, beyond its role in the structure of individual atoms and molecules. But this is a question of profound interest, because if entanglement is significant, then there is a direct physical sense in which a human being is a unity in significant ways, and not simply a collection of parts which each can be assigned their own individual properties.

So far we have described ways in which quantum physics may be supportive of traditional concepts associated with soul. However, we already

[4] C. H. Bennett, G. Brassard, C. Crépeau, R. Jozsa, A. Peres, and W. K. Wootters, 'Teleporting an Unknown Quantum State via Dual Classical and Einstein–Podolsky–Rosen Channels', *Phys. Rev. Lett.* **70** (1993), 1895–9; a non-specialist account is introduced in *A. Zeilinger, Dance of the Photons* (Farrar, Straus and Giroux, 2010), 46.

mentioned in Section 8.1 a third issue. This is that the many-worlds interpretation of quantum theory introduces a further radical change in which
the notions of value and disvalue, right and wrong are far from clear. From
a Christian perspective, one may feel that it is premature to reject the
many-worlds interpretation outright, but one would want to insist that
it not be allowed to undermine the *intuition to value*. That is, the human
intuition that *our acts are meaningful* should be allowed to be a true intuition.

8.3 Divine action

In the previous sections and chapters we referred to free will and responsibility, because these are at the heart of the question of human identity.
They also bear on the question of divine action in the world, because if the
basic nature of the physical world is such that it can furnish an ability of
conscious agents such as humans to comprehend their environment and
make freely willed responses, not just complex or random reactions to
stimuli, then there is in the world that richness of behaviour that is
affirmed in Christian and Jewish theology. That is, the world is not merely
the outworking of impersonal forces, but also the arena of personal
encounters and stories. If humans can enact personal encounters without
breaking out of the patterns we call 'laws of nature', then so can God.

Having said that, we have already explained that it is a mistake to
suppose that mere indeterminacy in physics is itself sufficient to resolve
the philosophical difficulties surrounding human responsibility. The
same is true of divine agency.

It is more helpful to note the elusive nature of quantum reality, and
from this deduce that the basic physics of the world may well have sufficient undiscovered richness to allow the correctness of the *intuition
to value* that we mentioned in the previous section. We said that this
intuition asserts that our own actions can be meaningful. We may add
that processes going on around us in the world can also be meaningful.
This is what the question of divine action is really about.

8.4 Some positive lessons to draw from quantum physics

We now review some of the lessons that quantum physics has for our
wider understanding, and for the life of faith. We list them under the

labels truth-seeking, beneficial technology, the Bible, unanswered questions, humility, biology, and metaphors.

Truth-seeking. Physics, like any other science, is about truth. It is about appreciating God's work by seeking the truth of it. It is part of what to love God with one's whole mind actually means in practice.

Beneficial technology. Science leads to technology. By understanding quantum physics better, we understand chemistry and materials science better, and we can design devices that, in good hands, lead to great benefits to people. Laser eye surgery, CAT scanners, and photovoltaic cells are just a few of the vast number of such beneficial devices. We express our mature and lively commitment to God by our mature, lively, and practical concern for our fellows. We bring sight to the blind by eye surgery, and improve health care by better diagnosis, and we seek to ameliorate human impact on the climate by finding better energy sources.

We are reminded of the parable of the sheep and the goats. This is the parable in which the scene of final judgement or calling to account is sketched by Jesus in brief powerful phrases.[5] In his telling, it is the people who actually did something for the needy who are commended, whether or not they did it for explicitly religious reasons. When we bring this together with other parts of the New Testament, the lesson is that the state of our hearts is more reliably indicated by our behaviour than by any supposedly loyal claims we might make.

We are reminded also of the parable of the two sons, in which a parent asked his sons to help him do a job, and one said he would help but did not, while the other said he would not help but in the event did.[6] In this parable it was, of course, the son who did the work who was commended, whether or not he said the right things. His action spoke louder than his words. In the story the work was in the father's wine business; it is up to us to interpret what that could mean for us. But it surely includes getting involved with making the conditions of life better for everyone, in so far as we can. The scientific method and the technology that it yields have been absolutely transformative in this regard. This is where God is showing us how to realize the wonderful potential that there is in the natural world.

The Bible. We pointed out in the previous chapter some examples of the fruitfulness of moving from form to function in thinking about the physical world. Rather than trying to say what quantum states or quantum

[5] Matthew 25:31–46.
[6] Matthew 21:28–32.

entanglement are, it has sometimes proved to be more useful to ask what they can do, or what can be done with them. As we commented in Chapter 5, it has in recent years become clear that a similar concern was present in the authors of the creation story in Genesis 1. In the light of other Ancient Near Eastern texts, and of the internal evidence, it can be concluded that the aim of Genesis 1 is not so much to describe the physical constitution of things, as to elucidate their role in the natural order.[7]

Unanswered Questions. Twentieth-century physics provides an object-lesson in how to live with unanswered questions. Sometimes religious life goes wrong when people agonize too much over questions they cannot answer, or when people try to provide answers prematurely. In the twentieth century, and right up to the present, physicists have had to accept that they have no clear consensus on the right way to interpret some of the basic aspects of quantum physics, but this has not prevented them getting on and making good progress. They have done this by developing good judgement on what are the most fruitful questions to ask next. The way to solve puzzles in science is often to seek more data. This optimistic seeking attitude is one aspect of the way of life that Jesus called the Kingdom of God.

Humility. In quantum physics, broadly speaking we know how to calculate what things will do, but we do not know what things are. Physicists and philosophers have bashed their heads against the puzzles of quantum reality for almost a century. One of the great lessons of this is epistemological humility. Epistemology is one of the core areas of philosophy. It is concerned with the nature, sources, and limits of knowledge. Quantum physics is an extraordinarily precise area of science, and we understand very well how to use it to calculate things like the magnetism of an electron and the energy levels of an atom to exquisite precision. The magnetism of the electron, for example, is expressed by a number which can be both measured and calculated (from quantum field theory) to a precision of a few parts in ten thousand billion (the value is 2.00231930436182). This is one of the most precise pieces of knowledge in the whole of science. However, we find it very hard to say exactly what a quantum state is, and therefore what physical things are, 'at bottom', as it were. We work around this by developing a number of different physical pictures, and respecting the mathematical rules of the subject, but we learn to rest

[7] J. H. Walton, *The Lost World of Genesis One: Ancient Cosmology and the Origins Debate* (Intervarsity Press, 2009).

content with the idea that we can't know everything. In particular, we cannot thoroughly know the nature of physical existence itself. Physical existence is a mystery, and quantum states are elusive. They have remained elusive for a hundred years despite very great efforts to understand them. Therefore, we acquire a certain amount of epistemological humility. We are not ashamed to say we do not know how even such a simple thing as a single electron can wave and move and have its being.

Biology. Is biology awaiting a quantum revolution? In one sense, quantum physics is already operational in biology, because it leads to the structure of all atoms and molecules, and the nature of liquid water, and so on. However, to a large extent the processes that are important in biology can be correctly understood without worrying about the quantum processes inside atoms. This is because the internal structure of each atom is very stable, and, to a large extent, atoms behave like separate little Lego blocks when they gather together to form larger things. The quantum states and entanglements are there, determining properties of single molecules, but the interaction between one large molecule and another, such as that between a protein and an enzyme, mostly does not depend on the subtleties that accompany quantum states. Therefore, to a large extent, biology can be said to be a study of complex processes that are, in principle, within the domain of classical physics rather than quantum physics.

However, the revolution that occurred in physics in the twentieth century makes modern-day physicists very aware of the danger of assuming one has the right basic concepts in any area of science. One may suspect that some such revolution will occur in biology too. It may be one in which quantum entanglement plays a role, or it may be another sort of revolution entirely.[8] What we learn from physics is that the rich success of one set of ideas does not rule out that that set of ideas must be eventually subsumed into something more subtle, in order to get a fuller understanding.

Metaphors. There are many great and important concepts of theology. They include the concept that God is love, and the concept that we can approach God in the way we would approach a loving father. A profound Christian concept is the Trinity, the idea that God's being is like a loving relationship or dance with a threefold nature. This has

---

[8] D. Abbott, P. C. W. Davies, A. K. Pati (eds), *Quantum Aspects of Life* (Imperial College Press, 2008).

proved to be a helpful way of expressing the richness of God while not leaving that richness completely vague or painted inappropriately. Another important lesson of theology is, however, that all attempts to capture God in our human expressions are doomed to fail in some way; they only act as hints and pointers, symbols or metaphors.

The theological idea of *Trinity* is and has always been an honest but tentative attempt to be helpful; that is what it remains. However, it has attracted scorn from sceptics, as if it is some sort of refusal to talk sense. Committed followers of Jesus have mostly felt that it does make sense, even if it can only be illustrated by imperfect illustrations. But the world does offer illustrations of the notion of 'three in one': the illustration of a triangle, for example, or, for the more mathematically minded, the notion of three dimensions. An illustration which we love is that of Borromean rings—three rings linked together in such a way that none intersects another, and yet they are bound together. Even better illustrations are to be found in loving human relationships, with their threefold sense of mutual giving, gift, and receiving. Artistic creation has similarly a threefold nature, which was insightfully explored by Dorothy Sayers: the act of creation, the work itself, and the act of reception in the mind of the one experiencing the work.[9]

The above examples serve either as symbols and metaphors for the nature of God or (when used unwisely) as insensitive barriers in the way of people's desire to learn.

Quantum entanglement offers us, if we can appreciate it correctly, another metaphor. Again, this is either helpful or unhelpful, depending on how it is used, and in either case it remains only a metaphor or a symbol or an analogy. It is no more able to capture fully the mystery of God than a triangle is. But a quantum state in the form of a *three-part entanglement* illustrates some of the themes that Trinitarian theology affirms. Such a three-part entanglement does not represent a property of God, but is a physical state of affairs that may help us appreciate God, somewhat as a religious symbol may.[10]

In a three-part quantum entangled state there are three physical entities that contribute some of their properties independently, but which also exhibit an inseparable nature, such that some properties of

---

[9] D. L. Sayers, *The Mind of the Maker* (Methuen, 1941); available at http://www.worldin-visible.com/library/dlsayers/mindofmaker/mind.c.htm

[10] J. C. Polkinghorne (ed.), *The Trinity and an Entangled World* (Erdmans, 2010).

the whole cannot be assigned individually to the parts, but exhibit purely a correlation or mutuality between the parts. Thus, the phenomenon illustrates, in a physical situation, the notion of 'distinctness yet inseparable togetherness' in a striking way. We consider this to be a helpful thought, but, once again, it should not be overstated. Among the positive benefits it offers is a reply to some, at least, of what sceptics have in the past said about what can and cannot make sense. Simple physical entities such as atoms and electrons can adopt states or configurations which might in the past have been said to be not just impossible but nonsensical. A quantum entangled state simply cannot be adequately described in the language of purely separate components, and if one insists on adopting the assumption that it can, then one will arrive at a contradiction. Quantum entanglement thus shows us that, even in the case of the natural physical world, the verb *to be* includes forms of existence that cannot be captured in simple notions of 'one thing next to another thing'. There is no shame, therefore, in allowing that God's nature can also be incapable of being captured in that kind of language.

Quantum entanglement does not, in and of itself, offer the number 'three'. The entanglement analogy has this in common with other analogies such as those of triangles and physical dimensions. There can be shapes with numbers of sides other than three, and spaces with a number of dimensions other than three, and entanglement among any number of parties. Thus, none of these analogies do the work of arriving at a threefold nature to God; that aspect is better understood in other terms, such as the more personal ones of loving relationships and self-communication, or that of 'beyond, alongside, and within'.

9

On the way

Andrew Steane gives a personal account

It has been my role to live much of life in an uncomfortable crossroads between conservative and liberal ways of being Christian, and between Christian and atheist ways of thinking. That is to say, having once started on the Christian way, I have always stayed on it, but I have been intensely aware of the objections raised by modern-day atheism when it challenges us to say what we have to offer and why it is better than the alternatives. *The way* or being *on the way*, on the journey, is one of the earliest expressions that the community of Christ used to describe itself in the first century, and I think it is a helpful one.

I was first persuaded of the basic truthfulness of the Christian message as a young man aged eighteen or nineteen. And that has always been what matters to me: *whether it is true or not*. Not whether it is helpful or comfortable. In fact I have found it to be difficult and not comfortable—but I feel the difficulty has been good for me. By 'the Christian message' here I mean the basic claim that we can learn from Jesus what God is like, and that our duty to acknowledge God becomes a way of life in which God gradually enables us to become the people we can best be.

I became convinced of the basic truthfulness of this by studying the Gospels and reading quite a few books, and talking it over with friends. One of the more telling points, for me, came from reading the philosopher Bertrand Russell's book *Why I am not a Christian*. Here was an intelligent and well-informed man telling me why he thought the message of the Gospels was either unworthy or incredible or both, and at first I went along with him. But as I began to think through his arguments it emerged that they were rather feeble. It was quite surprising to me that this capable philosopher would have written this rather feeble book.

One small example I remember was that Russell didn't like it when Jesus is said to have cursed a fig tree with the result that it withered. Russell thought there was something morally dubious about that tale.

If in fact it did not happen, then I have some sympathy for the view that in that case it should not have been written down as if it did happen. But Russell failed to address seriously the other possibilities, namely that the event did happen, or that its telling was a legitimate way to convey the lessons at work in it. If either of these possibilities hold, then we are dealing with a person of startling significance who could have every right to use a fig tree to convey a graphic, important, and memorable lesson.[1] So the issue is not, is it morally dubious? The issue is, are we here encountering such a person? Of course, one must answer 'almost certainly not' if this were an isolated incident in an otherwise unremarkable life. But, in fact, it is part of a larger story that gains more and more credibility and coherence the more one sticks with it. That, at least, was my experience.

Years later another statement of Russell's was to return to haunt me. This is his statement that Christian influence has been negative in the history of moral reforms:

> You find as you look around the world that every single bit of progress in humane feeling, every improvement in the criminal law, every step toward the diminution of war, every step toward better treatment of the coloured races, or every mitigation of slavery, every moral progress that there has been in the world, has been consistently opposed by the organized churches of the world. I say quite deliberately that the Christian religion, as organized in its churches, has been and still is the principal enemy of moral progress in the world.[2]

When I read that again after many years it gave me pause. Russell is no fool and his work has been hugely influential. Could it be that I had backed the wrong horse after all? It is not hard to find many examples of really unpleasant dogmas and attitudes in many churches, in either the past or present. If it is true that the Christian religion, as organized in its churches, has been and still is the principal enemy of moral progress in the world, then I had attached myself to something deeply ugly and it would be my duty to come to my senses and abandon it. Except

---

[1] The lesson concerns the fact that when a culture, or the practices and assumptions of a community, is no longer bearing fruit, then it is liable to be swept away. On this occasion Jesus himself swept away the coins on the tables of the money-changers at the heart of his own community.

[2] Bertrand Russell, *Why I am not a Christian and Other Essays on Religion and Related Subjects* (Simon and Schuster, 1957).

that my experience has not tallied with that. I have found the churches that I have joined to be largely good-hearted places making a genuine effort to help people less fortunate than themselves, and sensible about issues of politics and science and so on.

I mention this example because it conveys a true impression of my experience of trying to follow Christ, and of trying to be honest and serious about truth. This involves listening hard not just to people who are sympathetic to what you already think, but also to people who do not like what you think. I have found this to be very unsettling.

What I now think about this particular example is this. The Christian church has indeed got a terrible record in significant respects, and we must face up to this. However, it has also a great amount that it can claim on the positive side of the account, and in truth, Russell's statement is unjust and unbalanced. The truth is that, in the history of social reforms, Christian voices and commitments have been involved on *both* sides. The attitudes of Jesus have come into this situation and have acted like yeast, continually prodding the self-satisfied or powerful to open up a bit, and prompting reformers to take action, and encouraging the down-trodden. Russell was careful to say 'as organized in its churches' because, I suppose, he knew that there are plenty of examples of principled Christian people working for good. What his statement fails to do is take the trouble to get at the whole subject more insightfully. He contents himself with a sound bite, in what amounts to an attempt to generate unthinking attitudes.

In so far as they are large, bodies such as Churches are indeed hard to reform, but this is true of all large human institutions. Churches have, in fact, taken up and championed all the issues mentioned in Russell's statement. And if instead of Christian churches one looks at other identifiable groups of humans over a long period, one will also find a very poor record.

I had this kind of experience a second time when a relative asked me to read Richard Dawkins' book *The God Delusion*. This I did, and the experience was like being hit over the head repeatedly with a club. The book depressed me. To its credit, it contains a few brave and honest statements, objecting to injustice, but the second half contains large amounts of unsupported speculation in a style that verges on the pseudoscientific. However, when I first read it I did not want to dismiss it; I wanted to hear its criticism, and that criticism invited me to think of myself, and others like me, as a kind of second-rate human being, a deluded idiot who could not see plain facts about how rubbish the Bible is, someone ready to abuse their children with controlling fantasies having little more credibility than the tooth fairy.

It was only much later that I realized what had happened. I had been subjected to an intellectual mugging. The book adopts the tactics of propaganda and applies them with great force. There is disdain for history, elevation of obscure ill-argued studies, basic philosophical mis-understanding, and just enough truth to make an opening for all the rest.[3] But now my feeling of grief continues, not at the book directly, but at its reception. Large numbers of otherwise fair-minded and intel-ligent people think it is a good book. They did not notice that it contains not education but propaganda. This is a worrying state of affairs.

I shall mention two further components of my experiences as a fol-lower of Jesus. One is the way my understanding of what this involves has changed over time. The other is the way it has come into life decisions and professional work.

My earliest Christian teachers were in the Evangelical tradition. They were committed and energetic and in many ways kind-hearted, but they also had what now seems to me a rather narrow conception of what Jesus was talking about. For example, Jesus told a parable in which a farmer scatters seed, and a crop either grows or not, depending on the soil conditions. He then explained that the seed in the story represents an insight or 'word' from God and the different types of soil represent different types of human receptivity or the lack of it. In the Evangelical tradition I was taught that 'the word' here is 'the gospel', and 'the gos-pel' is the message that if you repent and believe that Jesus died for you then you will get new life with God. I now think that this is sort of half-right; it has some right ideas but the attitude of mind is too narrow, an attempt to contain life into a simple formula of words. I think now that Jesus was talking more generally about all sorts of ways in which God gives to us seeds of ideas which can liberate us, but we are not always receptive.

I learned from a book by Richard Rohr some very refreshing lessons in what Jesus actually stood for, and how it differs from what Christians have often said. For example, Jesus says that it is 'the peacemakers' who shall be called 'children of God', not the people who signed the correct doctrinal statement. He says that it is when we go the extra mile cheerfully

---

[3] John Cornwell, *Darwin's Angel* (Profile Books, 2007).

that we are being 'children of our true Father'; again, in this statement he does not require anyone to make any particular assertion
about the significance of Jesus' death, nor about his special significance
or role. The parable of the sheep and the goats makes it as plain as can
be that saying 'Lord, Lord' is not what he cares about (this is the part of
Jesus' account of final judgement that was referenced at the end of the
previous chapter; such judgement sets out a conclusive and absolutely
truthful statement of the reality of who we are). Finally, I think the
commission to 'make disciples of all the nations' does not mean 'convert everyone to conservative Christianity'; it means, 'encourage others
to join you on the journey of self-giving love and creative sympathy
and widening justice as we look to God, the source of these things.'

I found the poet R. S. Thomas helpful in my journey, because he also
is concerned about authenticity in our relationship with God, and he is
very honest about this. I found the pastor and writer Brian McLaren
helpful, and also George MacDonald, Simone Weil, Rowan Williams,
and others who are sensitive to the contemplative as well as other forms
of Christian action.

N. T. Wright has written insightfully about the fact that the hope
expressed in New Testament Christianity is not well captured by the
modern idiom of getting 'out of the world' and 'going to heaven'. It is
much more to do with transformation of the whole natural order, culminating in a New Creation which is not so much a replacement as a
blossoming, or something brought to birth.[4] This fits with the commitment to the world displayed by God in becoming incarnate in it. I found
this to be a liberating insight, and it prompted a study of the Hebrew
word *shamayim*. This word is commonly translated as 'heaven' or 'heavens', but this has become, arguably, a poor translation, because the
word 'heaven' has taken on a life of its own in the English language, and
it does not necessarily correspond to the Hebrew concept, or to what
Jesus may have been talking about. I don't think the Hebrew concept is,
in any simple sense, a place in a three-tier cosmos (Wikipedia is quite
wrong about this). It served rather as a way of talking about metaphysical
matters. *Shamayim* is a way of speaking about that which is absolutely
true and unbreakable; it is 'the place of God' in the sense of 'bedrock'
and in the sense of 'what can honourably and rightly be aspired to', so
it is also a way of speaking of moral values such as justice, liberty, fair

---

[4] N. T. Wright, *Surprised by Hope* (HarperOne, 2008).

distribution of resources, truthfulness, and the like. By treasuring these we locate our heart in the right place, as Jesus put it. He did not mean we should long for escape into a cushy paradise. However, he did add a note of hope: our present pain will not have the last word, and although we don't expect a cushy paradise, we can expect a festival.

My professional life has been based in a university—Oxford University—which, like many universities, has Christian origins but is now in an uneasy relationship with those origins. Academia has not known what to 'do' with religion for a long time now, and the result is that it has become almost a taboo subject amongst the academic staff. Everyone gets on with their particular academic discipline, and one must tread carefully if a conversation at lunch turns to anything which might be termed religious.

Of course, it is important to keep in mind that there is a sense in which following Jesus is an antidote to religion, rather than an example of religion. It is certainly not 'a religion', but this is what it is commonly thought to be. That is a convenient box to put it in if you want to keep it at arm's length. And, to be fair, there are plenty of muddled churches which have turned it into 'a religion', in the sense of a set of things to be believed and words to say, maintained by repetition in an inward-looking group. So I have found myself continually trying to negotiate this strange mix of distorted assumptions and honest efforts to do better.

The purpose of this chapter has been to write autobiographically, in order to share with the reader some human experience which might, we hope, be helpful to know. I am aware that in the above I have described a rather unsettled mixture of things, without much of a thread. I have written that way because I think it gives a true impression of much of my experience of being a follower of Jesus. But the picture would not be complete without adding the following aspects.

One of my parents was a very forceful person with severe mental health struggles. Somehow or other I seem to have found what was lost to me there. Not found in God; that does not quite capture the experience. It is more like: found in the communal way of life of God; the Kingdom of God. Like discovering treasure in a field. It has also been my experience that the most important decisions in life—decisions about marriage, and children, and the use of money, and keeping family connections alive, and priorities at work—have been greatly helped by being part of a Christian community with a treasure house of wisdom to draw from.

One of the most meaningful aspects of my life has been the decision to listen to and learn long-tested Christian wisdom, especially about what marriage is. I was encouraged to see singleness as a positive basis on which to serve others, and marriage as another positive basis by which two people can liberate each other to serve. I was also encouraged to see people more fully, including the deep links which extend to their other friends and family. In this way of seeing, none of our relationships are only about two people; they are all connections within the wider community.

It has twice happened to me that, before receiving something valuable, I had reached the point where I was willing to go without it, even though I still deeply desired it. The first was a soul mate, the second was children. I don't wish for a moment to suggest that this is the universal pattern, but I do think that in both cases, the willingness to go without enabled the relationship, when it came, to start on a good basis, one which could be more liberating for all parties. It is important not to push anyone into a role they cannot fulfil, and this is one of the ways in which we might inadvertently mistreat a spouse or a child. So I am grateful to God that I was spared the worst mistakes, and I think the Christian encouragement to be patient is one of the means by which this grace was given me.

Also, the decision to keep going to church, week in, week out, has had benefits which I did not realize would happen at the outset. One is that our children have been able to witness the life of the community, with its births and deaths, sufferings and celebrations. This has given them tremendous resources in their sense of who they are and how everyone has some role to play. I have also found it perfectly natural to give to my children space to become themselves without guilt-driven or controlling religious pressures. None of us is perfect but I don't think authentic Christian commitment need make one any worse at this than atheist or other commitment.

Finally, in my professional life I have had a journey from longing for acclaim to almost the opposite. As a young man I wanted to do something marvellous and have everyone say that it was marvellous. As an older man I want to do something to meet the needs of the time and place in which I find myself, and I am so far from caring whether anyone notices that I am almost wishing that it not be noticed. I don't quite go to that extreme, but I have become more and more aware of the amount of shouldering-for-recognition that goes on in academic life (as I suppose it also does in other walks of life), and I have longed for a different way of doing things.

Among the needs of the time and place in which I find myself, I would like to mention three items which require attention: first, global warming, climate change, and environmental destruction; secondly, global economic inequality (this feeds into the first); thirdly, the public misunderstanding of science and rationality. The third item risks being equally as damaging as the other items, and I would like to draw attention here to two elements. One is the idea that science reveals to us that the natural world is a vast machine and the 'scientific' conclusion is that there is no purpose or good or bad. This is a very great, and very common, misunderstanding of what science is and how in contributes. I have therefore tried to combat it.

There is also a widespread assumption that reason and rationality are squarely on the side of atheism, and to recognize God is an alternative to being reasonable, a decision involving some compromise of reason, or a desire to over-ride reason by 'faith'. This is so much assumed that it is taken for granted quite unquestioningly, in large parts of both the well-educated and ill-educated population. This unthinking assumption is like the assumption in eighteenth-century 'enlightened' conversation that the love shown by black people for one another was not genuine human love, but some sort of animal instinct. That is to say, it is a great prejudice and a great injustice. The widespread assumption that to acknowledge God is to be compromised in intellect, or to display something less than genuine rationality, is one of the great injustices of our time. It will one day be seen to be such.

10

General relativity, language, and learning

It is a basic principle in theology that human speech about God never fully and precisely captures the truth we are reaching for, and often must be largely metaphorical or symbolic or poetic. When we say that God is our Father, for example, we are making an appeal to a concept of parenthood that is not exactly the same as the one that applies in human relations, but which has a sufficient resonance with it that we find the thought a helpful one. The word 'Father' is here being used in a way that is hard to classify: it has aspects of metaphor and symbol and poetry and prose all at once.

In this chapter we shall present an extended example of a concept that emerges from an area of modern science, and that may present metaphors or analogies useful to theology, or just strange yet illuminating ways of thinking for the ordinary person. Note, the fact that this example comes from a well-established scientific field does not, in itself, determine whether it succeeds as a metaphor in theology. Nor should anyone think that by using science in this way we are making the mistake of supposing that theology gains better credentials when it is given a more 'sciency' feel. Indeed, we strongly repudiate such a supposition. However, the process of scientific discovery can and does furnish not just new information about the natural world, but also new concepts in terms of which we can learn to think. These concepts certainly help us to think more insightfully about the physical world. Some of them may also help us to think more richly about the nature of personal relationships, and hence about each other. From there we can reach more fully for that which is beyond us but which helps us to grow.

The subject we shall consider is the theory of space and time and gravity that is called general relativity and that was discovered largely by Albert Einstein, building on brilliant mathematical work by the German mathematicians Carl Friedrich Gauss and Bernhard Riemann.

General relativity is both a mathematical framework and a statement about the physical world. As a mathematical framework, it furnishes a set of ideas and equations that can be used to describe the notion of a 'space', and geometrical relationships within that space. As a statement about the physical world, it asserts that space and time together form an example of the kind of space that the theory treats, and it further asserts that physical objects themselves introduce changes in that space, warping it somewhat as cloth is warped when some of the threads pull tighter, and others loosen.

10.1 Map-making without privileged coordinates

A good insight into general relativity can be obtained by applying it to something simpler than the space and time of the physical universe. We shall, for our present purpose, not include time in the first instance, but just study space, and furthermore we will not worry about full three-dimensional space, but restrict our study to surfaces. For example, we might be interested in the surface of a sphere, or the surface of a flat piece of paper, or the surface of planet Earth.

Consider now the sort of atlas of the countries and geography of the Earth that one can obtain in a good bookshop. Such an atlas consists of a set of maps, and associated text and other pictures and diagrams. It is well known that it is not straightforward to compile an atlas of the globe, because, among other things, there is the difficulty of making a flat map of an object which is essentially spherical. A map of a small part of the surface of the Earth, say fifty miles by fifty miles across, does not face this difficulty, because such a small part of the surface is flat to good approximation. But a map, on flat paper, of the whole continent of Africa or Asia never altogether succeeds. It can show a lot of useful information, but it will never reproduce all the distances and directions in a direct mapping onto the flat page because this simply cannot be done.

To handle this, cartographers carefully provide 'coordinate' lines on the map, such as lines of longitude and latitude, and they provide a scale, and they make it clear that the scale of the map may vary from one place to another. For example, if at the centre of the map the scale is one centimetre to 100 kilometres, it may be that the scale varies smoothly across the map until at the edge it is one centimetre to

200 kilometres. The purchaser of such a map might ask the bookseller if she has another one without this awkward feature of a varying scale, but the bookseller will reply that if the scale is not allowed to vary in this way, then the map will not reproduce other features correctly, such as the angles between different rivers where they meet.

Overall, then, a flat map of a non-flat object must adopt a strategy to do the best job it can while accepting that it must mislead the user at first glance, if he or she does not pay attention to the scale or other such information.

Now suppose that a space probe is sent out to make a survey of an astronomical object such as a meteor. The overall shape of the meteor may be quite far from spherical, and it may have all sorts of mountains and valleys.

Suppose the survey is done, and you are given a set of maps of the surface. However, the maps are in the first instance furnished in a purely mathematical way, with no labels to indicate features. It doesn't say things like 'Tall Mountain' or 'Long Valley'. All you see is a rectangular mesh of coordinate lines (Figure 10.1). The map seems to show nothing whatsoever. At first you might conclude that the meteor is utterly featureless and flat. But then you notice that the scale of your map is not everywhere the same. The coordinate lines form a simple square mesh, but the readout from the probe says that in the middle of the map, one

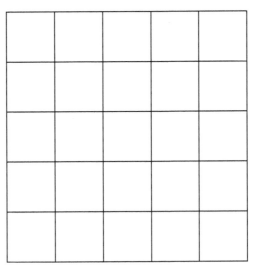

Figure 10.1 A map with coordinate lines shown.

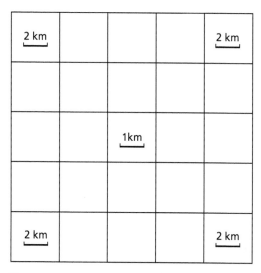

Figure 10.2 The same map as in Figure 10.1, but with scale information added. Note the grid squares do not all show the same area on the ground.

centimetre on the page corresponds to one kilometre on the meteor, whereas at the edge of the map one centimetre on the page corresponds to two kilometres on the meteor (Figure 10.2).

Puzzling over this, you might begin to see that this information about the scale is, in fact, telling you something about the overall shape of the meteor. The information you want—the information about the shape of the meteor—has been provided, but in an indirect way. If the scale is varying as we just described, then the interpretation is that the surface of the meteor is sloping more and more away from you, as you view the outer part of the map (Figure 10.3). Judging from the scale information alone, the slope might equally be towards you, but if you bring in the general notion that the map shows the outer surface of a three-dimensional object, then you will realize that the outer part of the map shows a region curving away from you, and furthermore the amount of change in the scale allows you to determine the amount by which the surface is curving. In technical terms, we say that the variation of the scale from place to place allows one to determine the radius of curvature of the surface. This is what Bernhard Riemann worked out in quantitative detail. In a detailed treatment, advanced mathematical methods are required. All we need here is the general concept.

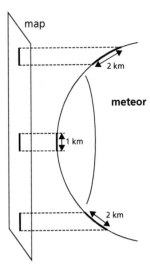

Figure 10.3 The scale on the map shown in Figure 10.2 implies that the meteor has a curved surface and the map is showing what is called a 'projection' of that surface.

The overall shape of the surface of the meteor is indicated by the overall change in scale of our map between the middle and outer part of the map. The scale can also indicate further information, such as local topographical features. Suppose, for example, that the grid lines remain everywhere straight and square, as in Figures 10.1 and 10.2, but there is a region where the scale locally becomes a lot smaller. Such a region indicates either a hill or a depression on the surface of the meteor. Similarly, ridges and valleys leave their trace by influencing the scale of the map. Slowly but surely, a complete picture emerges. One way to manage and automate this process of interpretation is to perform a lengthy calculation which translates from the scale information at every region to a statement of the deduced local curvature of the meteor surface. From this one can then derive a set of contour lines, and the structure of the meteor becomes clear. (The kind of map we have discussed is called a 'projection' or 'projective map'.)

So far we have described a mathematical idea which is not too far removed from everyday life, and therefore the analogies that it may suggest for theology are ones that are already part of standard discourse in theology. For example, it is already a well-known maxim, taught to theology students, that 'the map is not the territory'. The analogy we want to present has not yet appeared; it will emerge after the next part of our discussion.

Let us now suppose that another probe (Probe B) is sent out to survey an astronomical object, and the object in question may or may not be the same meteor as the one surveyed by the first probe (Probe A).

Probe A was designed in Israel, but this second probe (Probe B) has been designed in India, and it returns its information using a different coordinate system. This time the coordinate lines appear on the map as a set of lines out from the centre, and a set of concentric circles (Figure 10.4).

Figure 10.4 only shows the coordinate lines. Since this figure does not show any labels or features, and no scale, it does not, as it stands, furnish any information about the surface of the astronomical object being surveyed. However, once we add the scale information provided by the probe, we will be able to begin to interpret the map.

We imagine that the space probe sends its information back in the form not of a picture but of a data stream consisting of a set of numbers. The numbers give the scale of the map at every coordinate location. By bringing together all this information in a suitable calculation, one can once again determine the shape of the astronomical object that has been surveyed. (In fact, real space probes do not typically work like this, but this extended example correctly shows the nature of the mathematical structure of general relativity in the way it is most commonly used to survey space–time.)

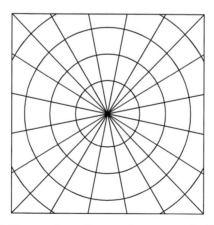

Figure 10.4 The coordinate lines on the map supplied by probe B.

Now we will draw out some of the striking features of this approach to mapping, and some of the pitfalls that lie in the way of understanding.

The first striking observation is that although Figures 10.1 and 10.4 look very different, it is perfectly possible that they may be unlabelled maps of precisely the same surface. Because these figures only show the coordinate lines, they do not in fact show any information at all about the astronomical object in question. In a rough analogy, the coordinate lines are like a set of questions, but until we have the answers to the questions, we have not yet learned anything (except perhaps the skill involved in asking good questions). An unwary viewer who saw Figures 10.1 and 10.4 in an atlas might wrongly 'deduce' that two totally different places were on show, but in fact there is no way of telling until we have further information. Note that this is not about looking at the same object from different directions. It is about preparing to survey any object using different coordinate systems.

The next striking observation is one we have not yet indicated in our discussion of these different coordinate systems. It is a remaining ambiguity that is an essential feature of this area of mathematics and geometry. The problem is: when we prepare the coordinate systems *before* surveying the astronomical object, we have no way of knowing whether the coordinate lines or our map are going to correspond to straight lines on the ground. In fact, they almost certainly will not. In the case of a survey of the surface of a three-dimensional object, this difficulty is not too disturbing, but more generally this can present very great problems of interpretation. A perfect probe will provide the correct scale information in terms of the coordinates it has been required to use, but the resulting data will be almost impossible to interpret if the coordinates were poorly chosen.

The technical term for the scale factor in general relativity is the *metric*. The word has the sense of 'that which measures' or 'that which shows the measure'. In the discussion of space-time, the metric contains ten different numbers or 'scale factors', each of which may vary from one region of space-time to another, and they may have either sign (positive or negative). Roughly speaking, if those of positive sign indicate separation in space, then those of negative sign indicate separation in time. The form or 'shape' of space-time is indicated by the way these numbers vary from one location in space-time to another, and by their signs.

The difficulty in interpreting what the metric is showing is because poorly chosen coordinates may wander about space-time in unknown ways, and one has no way of knowing what is happening until the metric is obtained in terms of those coordinates. Furthermore, one may find that what one thought was a distance in space turns out to be an interval of time, or vice versa.

Examples of this have regularly occurred in the study of general relativity over the past century. Indeed, the problem is so difficult that only a few simple scenarios have been analysed in full. One example problem is what happens at the surface of a region of space that has come to be known as a 'black hole'. For several decades, physicists were not sure how to interpret what the equations of general relativity were telling them. From one point of view, time itself seems to slow to a stop at the surface or 'horizon' of a black hole, and for this reason physicists thought that such a horizon could only form infinitely slowly and therefore would never fully form. However, from another point of view (that is, in terms of another system of coordinates), time does not slow and it seems that a black hole horizon can readily form in a short time when the conditions are right.

Eventually a somewhat tentative consensus was arrived at that certain physical processes could result in a black hole. This was sufficient to promote a search, but it was with a real sense of uncertainty that astronomers looked for evidence of these objects in the universe at large. They had a complete and essentially correct theory of space-time to guide them, but that very theory was hard to interpret with confidence. There is now a large body of evidence for the existence and nature of black holes, and they have become a standard part of the astrophysical lexicon.

Another example of the difficulty of interpretation in general relativity is the phenomenon of gravitational waves. The equations of general relativity imply that there can be regions of space which oscillate in a sort of 'stretch and squeeze' motion. However, it was not clear whether these were merely oscillations in the coordinates, with space remaining still, or genuine oscillations in the underlying space. It is as if some pages of an atlas show a vibrating image, but one needs to look carefully at the scale information to determine whether one is viewing a fixed terrain through a vibrating lens, or a vibrating terrain through a fixed lens. Here the 'lens' is furnished by the coordinate system, and it is only with difficulty that one can determine that these waves are genuine disturbances in space. Furthermore, even after arriving at a firm view on that, there remains the further issue of whether these disturbances can be detected by physical detectors. Again, it was not immediately clear because if the very ruler one is using is also being stretched, then what exactly does the stretching mean in terms of observable consequences?

It turns out that these vibrations in space are genuine and they can be detected by sufficiently large and sensitive seismometers. Note, here we are not discussing a vibration only of the ground of a planet such as Earth. We are discussing a vibration of the very space which holds and surrounds the Earth. This is a very counterintuitive concept, and the detection requires extremely sensitive optical instruments with light beam paths extending over several kilometres. The first direct detection of a gravitational wave was achieved in September 2015 by the LIGO and Virgo scientific collaboration. This was a marvellous and celebrated culmination of decades of effort by a team of many hundreds of people from many different countries. The 2017 Nobel Prize in Physics was awarded for this brilliant work.

10.2 Theological reflection

All that we discussed in Section 10.1 have lessons to teach about the difficulties of theological reflection. Some of these lessons are obvious, but some are quite subtle and these are in many respects more telling.

Before we discuss such lessons, we would like to make clear our view that theological reflection is not done appropriately, or with any prospect of progress, when it is undertaken in a purely passive way, with no commitment on the part of the reflective person to be responsive to what they find. So an analogy with surveying an astronomical object using a passive probe fails severely at this point. Our goal is not to survey an object, but to know the very convener of all our meetings. However, the subtle interplay of coordinate and scale information fundamental to the discipline of general relativity does have useful wider lessons to offer.

The first lesson is quite close to the well-known one that we are only able to receive truth when it is offered to us in a language that we can understand. Our geometrical example already illustrates the point that, even in the context of something as simple as the shape of a physical object, the language we adopt may itself be so configured that it is either easy or difficult for a given truth to be expressed in it. General relativity makes this fact even more telling, because in general relativity one does not even know, at the outset, whether a change in a given coordinate represents some time passing or a separation in space or a combination of both, as we already mentioned. So the question 'How far is it from A to B?' might have the answer 'one week', and this might mean not a week-long journey, but rather that A and B are coordinate labels for the very same spatial location, marked at different times. Perhaps this is not too disturbing, but consider the question 'Is it further from C to D or from A to B?' This seems like a fair question, but it might be muddled, because CD might be a spatial distance whereas AB is a time, and the point is that the coordinates you are using will not themselves allow you to know this. It is the scale information coming back from the probe or *metric* which tells you the truth of what your coordinates really mean.

It is a well-known phenomenon of language that the same question may have different answers depending on the framework or set of assumptions that have been adopted, but we are making a stronger point. In general relativity, and in theology, it may often be that *one does not even know, at the outset, what type of question one has oneself asked*, but this does not make the question meaningless nor an answer impossible. But it may happen that the answer comes in a form for which one was unprepared, and this is not owing to a perverse desire to be obscure, but from a pure intention to be as truthful and illuminating as possible.

Also, it is only in a long sequence of answers to a variety of questions that the shape of the terrain begins to be possible to discern, and this shape will itself tell us what is a better way of framing our questions.

Imagine an ancient Hebrew soldier asking God, 'What kind of a tribal god are You?' What answer can be given? One answer is 'No kind. I am not a tribal god.' Another answer is 'The kind that requires you to be fair and just to your fellow Israelites.' Another answer is 'The kind that wishes you to become the sort of person who would lay down your life on behalf of the very Canaanites you are currently so eager to destroy.' The first answer is, we may reverently and appropriately hold, largely true, but it does not help the Hebrew soldier very much. The second answer is less than the whole truth but is perhaps helpful in the circumstances. The third answer is, from our Christian perspective, profoundly truthful, but it is probably so alien to the thought-space of the ancient Hebrew soldier that it simply would not be able to be maintained in his thought processes. The idea of triumphant martial victory over enemies is very widespread in the history of the human race; it is practically universal. We should not look down on that soldier, but rather be thankful for the real moral and spiritual progress that has been made since those ancient times, and play our part in maintaining and further illuminating this most precious inheritance.

Now we would like to come to a more difficult application. This concerns an issue close to the heart of what many people find to be an intensely precious discovery, namely the personal nature of God.

It is, we believe, important to make a balance between eagerness to make assertions about God and caution not to say something misleading. In theology the terms *cataphatic* and *apophatic* are used to refer to the two approaches. The first says positive things, such as 'God is like …'. The second says negative things, such as 'God is not like …', often in order to emphasize that even the best words we use may be inadequate in one respect or another. Both approaches are needed. It is with this in mind that one may truly say that 'God both is and is not willing to be called personal.' This statement might feel threatening to some readers, because if misunderstood it might look like a compromise between faith and unfaith. What we mean by it is that when we adopt the language of personhood in our approach to God, this is appropriate, but no human concept is adequate to capture in any brief phrase what can be said of God in complete truth. Our concept of personhood is a valuable 'coordinate' which, in company with other 'coordinates', can be used by God to convey to us a growing appreciation of God's own nature, but that nature is not captured by any one of our phrases.

A friend of one of us (AS) once asked, in the context of a small prayer group, 'How do we know that God loves us?' and this was a genuine question to which trite answers were not required. The person in question had undergone and was undergoing deeply painful experiences of oppressive family life and severe mental illness in a loved one. The atheist has a ready answer to this question, which is that the whole of our existence is founded in impersonal principles and there is no love whatsoever, or even the capacity to either care or not care, outside of complex organisms such as human beings. The follower of Jesus judges, on the basis of a large number of considerations, that God both can and does love, but it is deeply challenging to translate what this means directly into everyday human situations, because we ourselves are the words of that translation. A valid answer to our friend's question is 'Because otherwise that one would not be God.' In other words, the possibilities are either atheism or that God does love us. (The line you sometimes hear, that if there were God then God would not care about individual people, is quite wrong; such a one would not warrant the name 'God'.) This is far from meaningless, but this answer in words alone is rendered meaningless in a given human setting if the person or community offering it does not live accordingly; i.e. perform the loving actions that would be performed by one who truly acknowledged God.

We, the authors, consider that someone who responds to this question by a supportive action while in words honestly denying the reality of God is a truer servant of God than one who does nothing but gives a doctrinally correct reply. A father once asked his two sons for help, and one expressed willingness but did not in fact do anything, while the other said he would not but then helped anyway. We already observed (in Chapter 8) which son pleased his father more.[1]

We conclude this chapter with a brief reflection on why we introduced the example of general relativity in order to convey some ideas about language and learning. The reason is that it is the intense experience of anyone wrestling with general relativity that the subject matter remains always at one step removed from one's immediate grasp. The subject matter is space–time, but we never experience space–time directly, we only experience things that move and interact in the context of space–time. The mathematical framework of the subject does not give us a 'handle' on space–time directly. It offers us ways to connect one part of any given map to another, and to compare maps, but it does all this without ever saying exactly what it is that we are mapping. If such an experience can occur even in such a comparatively simple area as the measurement of distance and time, then how much more will it occur in the understanding of personhood and personal relations! It should not surprise us that the nature of God's personal being is so mysterious that it shocks many people into denying it altogether. This should not be interpreted as pure wilfulness on the part of our fellow citizens. It is largely driven by pure puzzlement as to what we could possibly mean. We must do our best to explain ourselves—to set out our vision of truth in a way that others can deem worthy of attention—but in the end we show what it can possibly mean chiefly by the way we live.

It is Christian actions, not Christian doctrines, which illuminate the world (when they are truly Christ-like and not an abuse of that name). However, those actions need a wellspring to provide the perseverance

[1] Matthew 21:28–32.

and sense of adventure that is required, and that wellspring consists, in part, of the Christian vision of life. This is a vision in which human life itself serves as a set of coordinates through which beautiful truths of existence can be expressed. In an analogy that is loose and imperfect, but nonetheless we hope helpful, Jesus serves the role of 'metric'. He shows what form divine *being* takes when it is expressed in the coordinates of human being. The value of this analogy lies in the fact that it captures the idea that the truth of things is beyond our ability to express or even see directly, but we can see what form that truth takes when it is expressed in the sort of life that human life is.

Statements about divinity are always hovering at the boundaries of language; we are trying to speak of things we hardly know, and this is similar to what one must do when trying to speak precisely about the arena that we call space–time. The difference, though, remains very large. In the second case one can and should try to express the object of study in mathematical equations. In the first case one is not dealing with an object of study but a partner in a relationship.

11

The argument from design

11.1 Introduction

The American astronaut John Glenn was a deeply religious man. When he returned from his final mission to space he said, 'To look out at this kind of creation and not believe in God is to me impossible.' I (HH) have a similar attitude—when watching BBC's *Planet Earth*, or when I first saw my newborn son, or when I first understood the connection between gravity and space–time curvature in Einstein's theory of relativity. When I experience things like this, I find myself powerfully and naturally led to the idea of an intelligence far greater than my own.

I won't say that belief in God is *irresistible*—for many intelligent people find a way to resist it. But if my case is at all typical, then curious people might find themselves with an inclination to believe in transcendence, in some creative intelligence lying behind the blooming, buzzing confusion of our daily lives.

If you're concerned also with the rationality of religious belief, then there's a temptation to take the natural impulse toward the transcendent and to try to transform it into a 'scientific' argument for the existence of God. Many people throughout history have succumbed to this temptation—perhaps Thomas Aquinas, for example (though we will question this shortly), and certainly many of the philosophical commentators. In its simplest and most naive form, such an argument might run like this:

> There are aspects of nature that are too beautiful (or too complex, or too purposeful) to admit any naturalistic explanation. Thus, the best explanation for these aspects of nature is the existence of an omnipotent creator.

In a moment we are going to comment on this from a scientific point of view. But before we do, it is important to point out a fundamental difficulty with this kind of argument. The problem is that the phrase at the end, 'omnipotent creator', appears to refer to a being that one might or might not consider as worthy of one's allegiance, and if that is really what the phrase refers to, then the phrase does not refer to God. In fact, anything so paltry that its very existence could be questionable is not God and should not be named as God.

Many works of philosophy, and of Christian apologetics, have been presented as 'arguments for and against the existence of God'. Such arguments are the symptom of an attitude that has already gone wrong. It is an attitude that puts God at our disposal. The whole mindset adopted in this type of argument is misconceived, because it puts ourselves and our arguments in the role of final judge and arbiter, and this is not a role we can or should try to claim. The situation is subtle, though, because in some sense we can't help being the judge over our own decisions. The way out of this dilemma is not for us to stand in judgement over God, as if we might or might not deign to allow that God exists. Rather, it is our role to take on a seeking mindset, one that is ready to allow our decision-making to become more and more in tune with whatever are the right attitudes—the attitudes that *God* stands for. We name as *God* that which is real and worthy of our allegiance and make it our aim to *learn the nature of that One*. We tentatively develop a sense of connection to God as we consider that God has made Godself known to us, as, for example, happens in our experiences of transcendence, and in our learning of what happened in first-century Palestine.

For clarity, let us repeat: we want to refuse, right at the outset, even the suggestion that one should first define some X, then give arguments by which the existence of X is derived, and then acknowledge X to be God.[1] Instead of this, we are asserting that the right approach is simply to take on a willing attitude: willing to learn what deserves our allegiance, and willing to give that allegiance. In short, asking God to announce Godself to us. The first approach, where we make God the end point of a philosophical puzzle, places either ourselves or the argument in the role of primary judge or final arbiter, and that is the very role we must relinquish. In the second approach, we gather up the whole of our experience, making ourselves both humble and open to reason, and make it our aim to learn what it means to say that the universe is God's. This does not at all imply that we make no attempt to be rational in our approach to God. Far from it. The attempt to get the reasoning correct is the very thing we are doing, both in this chapter and throughout the book!

The work of Thomas Aquinas has often been presented as if Aquinas was trying to provide 'arguments for the existence of X' where X is God.

---

[1] Genesis 1, which we paraphrased in Section 5.5, avoids this mistaken approach.

If this is so, then he fell into the attitude which we are saying is wrong. However, that may not be what Aquinas was doing. He may have been pondering various ways of saying what we mean by God, or how to find one's way to God, rather than trying to offer arguments for the existence of anything whose existence was questionable (we already looked at this in Chapter 3). So perhaps Aquinas was more theologically wise and sensitive than some of his commentators have given him credit for.

We have now made it clear that the style of argument which we have begun to address in this chapter is questionable from a theological and spiritual point of view (by 'spiritual' here we mean *concerned with attitudes and with the whole human person*). We will now show that it fails from a scientific point of view as well. Consider again the example argument from aspects of nature given earlier (third paragraph of this section). How does this argument fare as a scientific proof of the existence of a being with attributes like God's attributes? According to many bright minds, this argument is an abject failure. Already in the eighteenth century, David Hume attempted to eviscerate the design argument—and was largely successful. Hume's criticisms were taken on by Immanuel Kant, who has had an enormous influence on philosophical thought up to the present day. In fact, some theologians believe that Kant put the final nail in the coffin of natural theology. And we don't even need to look to abstruse philosophical arguments to uncover problems with the design argument, for Darwin's theory of evolution already fills many of the gaps that have been cited in traditional design arguments.

What then are the other scientific problems with the design argument? Often the argument is put forward in an imprecise fashion, making it difficult to evaluate by scientific standards. Science doesn't speak of quantities of 'beauty', and even the notion of 'complexity' is plagued with scientific difficulties. Another problem with this argument is that it leaps quickly to the conclusion that these aspects of nature do not— or even cannot—admit of a naturalistic explanation. But who knows what kind of resources future science might have for explaining things? And what's more, who really understands the supposed boundary between 'natural' and 'supernatural' explanations? After all, the idea of an electromagnetic field might well have been considered a supernatural explanation to people living before the nineteenth century.[2]

---

[2] In the Middle Ages, the Latin equivalents meant the following: natural = what God does all the time; supernatural = what God does on special occasions. The later deist philosophies (which we introduced in Section 3.3) redefined this as follows: natural = the way things work without God; supernatural = God intervening from outside. Since

In short, if the design argument is psychologically forceful, it is also notoriously difficult to make rigorous. What are we to make of these dual features of design arguments? In our opinion, the psychological force of design arguments must be taken seriously. In fact, we think that it is perfectly natural to approach the world with a sense of wonder so strong that it points beyond the natural world and to its creator. Nonetheless, we think that human reason goes astray when one tries to transform this sense of wonder into a rigorous scientific confirmation of the existence of God. We will explain these points in detail in the pages that follow. But for now, let's focus our attention on the most recent incarnation of the design argument, the so-called cosmological fine-tuning argument.

11.2 Cosmological fine tuning

In the early twentieth century, cosmology took a turn toward scientific rigour. After Einstein discovered the general theory of relativity, he realized that the theory admitted cosmological models—models representing the entire large-scale history of the universe. That is to say, the equations that describe the behaviour of space and time can be used to make precise statements about the possible ways in which the cosmos can develop. Einstein's first proposed cosmological model was a static universe. This later proved to be inconsistent with the observed red shift of distant galaxies. An alternative model, in which the universe is finitely old, was proposed originally by Georges Lemaitre. Such 'finite big-bang' models were later developed by a number of physicists such as Friedmann, Robertson, and Walker.[3]

A key feature of Einstein's theory of general relativity is that it has two components. The first component is Einstein's field equations, which represent the laws of nature. These laws hold true in any possible universe that embodies general relativity. (Here we use the word 'possible' in the narrow sense of possibility within our current scientific understanding. Of course, there is a wider sense of logical or metaphysical possibility.) The second component of Einstein's theory is the selection

---

this redefinition is based on concepts which we consider to be flawed, we think that it leads to erroneous thinking, not least about miracles. But words have history, and it is probably not useful to try to recover the medieval meanings in their context.

[3] For a scholarly account of the development of physical cosmology in the twentieth century, see Helge Kragh, *Cosmology and Controversy* (Princeton University Press, 1999).

of some particular solution to Einstein's field equations. In this case, a particular solution to Einstein's field equations describes the main large-scale features of an entire universe. If Einstein's theory is true, then our universe is described (on the large scale) by one particular solution of the field equations. Other possible universes are described by other solutions of the field equations.

Some of these universes are similar to our own, and others of them are radically different. For example, the field equations are consistent with a universe that comes into existence for just a few seconds, and then contracts down into a Big Crunch. Such a universe would not have allowed for the development of sentient life in the way that our universe did.

We might wonder then: of all the initial conditions that we might specify for Einstein's field equations, how many of them could potentially have led to the development of sentient life? That's a difficult question to answer, especially since the concept of sentient life doesn't come up in Einstein's theory. However, what physicists have done is to come up with a set of minimal necessary physical conditions for the evolution of sentient life as we know it. As a shorthand, we will say that a universe is 'nice' if it satisfies these minimal necessary conditions for the development of life.

We can then raise a theoretical question: if we randomly pick a solution to Einstein's field equations, how likely is it that this solution would be nice? Well, physicists have tried to answer this question, and almost universally they agree that the nice universes are extremely rare. In fact, these nice universes are so rare that physicists such as Roger Penrose have said that the chances of choosing a nice universe are less likely than the chance of immediately finding a needle in a haystack that is bigger than the universe.

We now have the ingredients for the cosmological fine-tuning argument. Here's how the argument goes[4] (to be clear, we are *not* endorsing this argument):

> If God doesn't exist, then the chances of a nice universe are exceedingly low. If God does exist, then the chances of a nice universe are reasonably high. But the universe is nice. Therefore, it's likely that God exists.

---

[4] Recent versions of this style of argument can be found in Richard Swinburne, *The Existence of God* (Clarendon Press, 2004), and Robin Collins, 'The Teleological Argument: An Exploration of the Fine-tuning of the Cosmos', in W. L. Craig and J. P. Moreland (eds), *The Blackwell Companion to Natural Theology* (Blackwell, 2009).

This way of stating the argument involves some tacit assumptions. Someone positive about the argument, and proposing it, probably has in mind that they already know God in some measure, and the argument perhaps encourages them that they are on the right track. Someone who has no such sense of the reality of God, on the other hand, will treat the argument as a way of showing them that they have missed something, and they are liable to find it unpersuasive. They may judge that the argument is asking them to think along the following lines:

> We know that the universe is nice. When we discover that the chance of the universe being nice is extremely low, then we have a puzzle. Why is it this way? The best solution to this puzzle is that the reality that brought about the existence of the nice universe is able to select or tune the conditions by making a considered choice, somewhat in the same way as we are able to make considered choices.

They will then immediately want to ask: 'Is that the best solution? In what sense, "best"?'

Note, we are finding it hard to even present this argument without writing phrases that we consider to be meaningless or self-contradictory, such as the phrase, 'If God doesn't exist'. As we already pointed out, such a way of speaking has already failed to use language appropriately. It is like an argument that starts out from, 'If a circle is not everywhere equidistant from its centre, then …'. Anyone who begins an argument that way clearly either does not understand what a circle is or is using the word 'circle' to refer to some other thing, something which is not a circle. The same goes for arguments that speak of 'God' as something other than the very ground of being, or as other than one whose very name and nature is 'I am'.

Notwithstanding this muddle, the fine-tuning argument finishes with a statement we agree with, namely that God makes considered choices. (More precisely, there is a capacity enjoyed by God, of which our capacity to make considered choices is a partial reflection.) The logic here might be compared to the sequence

$$1 + 1 = 0,$$
$$2 + 2 = 6.$$

Therefore,

$$1 + 1 + 2 + 2 = 6.$$

In this thoroughly muddled sequence of arithmetic steps, the final line is correct, but this does not make the argument acceptable! In a similar way, the fine-tuning argument may finish in a statement which is true though unsupported by the argument. We are not saying that the problem with the fine-tuning argument is as obvious as the problem with this arithmetic argument. We introduced this example merely to make clear that to reject an argument does not necessarily mean that one holds its conclusion to be false.

It is notoriously difficult to critique an argument when you agree with its concluding statement! And yet we are going to do so, not just for the first and rather ill-stated version of the argument that we gave, but also for the second version. We agree that God is able to make considered choices, and that God created the heavens and the earth, and that God intended for there to be sentient beings. And yet, when we look at the logic of this argument, we find it wholly unpersuasive. There are a number of points on which this argument falters.

11.3 The infinity problem

First of all, notice how quickly we started talking about the chances of a nice universe. How are we supposed to figure out these chances? Well, surely Einstein's general theory of relativity is supposed to tell us, right? Alas, things aren't so simple. There are infinitely many different solutions to Einstein's field equations. How then are we supposed to count proportions of infinity? Unaided human reason doesn't provide clear intuitions about the chances of selecting a certain sort of thing from an infinite collection. Suppose, for the sake of argument, that it is a question of choosing from infinitely many possible universes, with infinitely many of them nice enough for life to develop, and infinitely many not. This does not tell us what the chances are that a random universe would be nice, because we do not know the relative likelihoods of the different parameter values—what statisticians call the distribution function.

Where, then, can we turn for some guidance on this question? Of course, we ought to turn here to empirical science itself. Now, empirical science does have something to say about the chances of various universes. Nonetheless, the arguments here are a bit speculative, and physicists have yet to reach universal consensus on these issues. In short, there are some plausible procedures for determining the chances of a nice universe, and these procedures depend in their details on currently accepted cosmological theory (namely, Einstein's theory of relativity). If Einstein's theory were to be replaced (as it doubtless will be), then there will be different recipes for calculating the possibility of nice universes. For all that we know, future cosmological theory may say that a nice universe was guaranteed by the laws of nature.

It has happened before that things that looked coincidental or specially arranged turned out to be better thought of another way. An example from physics is the fact that the inertial mass of a body, indicated by its tendency to exhibit inertia and resist acceleration, is equal to the gravitational mass, indicated by its tendency to attract other bodies by gravitational attraction. It is because of this that bodies released near to Earth all fall with the same acceleration (in the absence of air resistance). This seemed like a remarkable conspiracy, when seen from a Newtonian perspective, but with Einstein's treatment of gravitation it becomes simply the way things must be. The apparent 'conspiracy' of fine tuning is not quite like this, because in the gravity example there is an exact equality, whereas in fine tuning there is a very precise non-zero number, but it serves to illustrate that science can open up new perspectives on the issues that it reveals.

And now consider the following example.

The standard and largely correct model of any ordinary gas, such as air or steam, is that the gas is made up of a very large number of small molecules moving about randomly. If there is some gas in a large glass jar, for example, then the molecules are spread evenly through the jar. Therefore, at any given time, the probability that any one particular molecule is found in the bottom tenth of the jar is about one in ten. To find the probability that two particular molecules are both in the bottom tenth of the jar, one multiplies (it is like rolling a pair of ten-sided dice) $0.1 \times 0.1 = 0.01$, so the answer is one in a hundred. For three molecules it is one in a thousand, and so on. Note how quickly the probability falls as the number of chosen molecules increases. In a one-litre jar at ordinary temperature and pressure there are about 10^{22} molecules (that is, 1 followed by 22 zeros, which is ten thousand billion billion). The probability that all of these would be found all at once in the bottom tenth of the jar is, then, one-tenth multiplied by itself 10^{22} times, which gives a number so small that it is hard to describe how small it is: $0.000\ldots0001$ with not twenty-two zeros, but 10^{22} zeros before the 1. This is similar to the chances of winning the UK National Lottery one thousand billion billion times in a row. But now suppose that the gas in the jar was, in fact, largely water vapour (steam) at a temperature initially above 100°C. And suppose we simply lower the temperature at constant pressure. As the temperature passes below 100°C, the water vapour will start to condense into water, and gather at the bottom of the jar. This happens quite naturally, with no magic required, and no one needing to take any decisions. The water molecules will soon mostly be found in a small pool filling much less than even one-tenth of the jar. So, the event which seemed to be astronomically unlikely, in fact happens every day.

What this example illustrates is that using probability can be a poor way to deal with lack of knowledge, particularly when significant factors or mechanisms are not known. What happens when water condenses involves both the speed of movement and the location of the molecules, and it involves the attraction between them. When two molecules collide, or when they collide with the walls of the glass jar, they cling together and if their motional energy is not too great then this leads to a cumulative effect in which the molecules rapidly gather in larger and larger groups until they form a liquid. Our calculation in terms of probability did not take this into account, and as a result was misleading and essentially wrong.

This illustration should not be taken to imply that the conditions of the universe are unremarkable after all. It remains very puzzling why any natural process that can act at the level of basic conditions of the cosmos should happen to result in life-promoting conditions, a nice universe. But the illustration does suffice to show that an argument from fine tuning to the conclusion that an intelligently considered choice must have been required to do the tuning is, as a matter of logic, utterly unconvincing. Even if we knew what numbers to put into the calculations of the probability of the universe being suitable for life (we don't, but we are rapidly learning more), there would still remain the possibility of some unknown factor which we have failed to take into account.

It is by no means ruled out that progress in cosmology will come to give a perspective in which a nice universe is an altogether expected outcome of the way the universe works. Imagine for a moment that this did happen at some time in the future. Then we would find ourselves reflecting on a different state of our knowledge, one in which a universe configured for the emergence of life might be expected and unsurprising. Would you then find yourself inclined to argue that, in view of the arrangements of the cosmos being congenial to the development of life, then it follows that the situation must have been arranged by design—by a considered free choice made by God? But you can't have it both ways! Such an argument and the fine-tuning argument cannot both be right. But both can be wrong, and indeed both are wrong, as we now propose to show.

11.4 Does X = God?

Traditional design arguments came under heavy attack in David Hume's *Dialogues Concerning Natural Religion*. While some of Hume's criticisms might have been unfair, he correctly points out that design arguments always fall short of establishing that God is as the Abrahamic tradition says God is. That is, while design arguments might (if successful) prove the existence of some thing X that caused the existence of our universe, still they don't provide evidence for the existence of One capable of caring about anyone, or Who loves.

Now, many advocates of natural theology have replied to Hume's complaint by saying that the design arguments aren't intended to *demonstrate* the existence of God as God is described in the Bible. Rather, these arguments are intended to be suggestive—to point us in the right direction. But there remains a worry in our minds about the methodology of fine-tuning arguments: these arguments require us to theorize about God. Now, if we take the word 'theorize' in its loosest sense, as 'thinking about', then a religious believer cannot help but to theorize about God. But there is a more technical sense of 'theorize' that carries constructive overtones. For example, when scientists theorize about the inner structure of the nucleus of an atom, then they 'make up a story' about what's going on in the atom—and then they try to gather evidence about whether this story is correct.

Is it acceptable to 'make up a story' about who God is and how He acts? It's at this point that we start to worry, and for multiple reasons. For one, the prophet Isaiah reports God as saying, 'For my thoughts are not your thoughts, neither are your ways my ways.' So, we worry that any story we come up with about 'what God would do' is likely to be dead wrong. Related to this first worry, there is another worry having to do with the nature of God. Many Western monotheists find that God has attributes such as thoughtfulness. To be more precise, God has a quality of which our own thoughtfulness is a pale reflection. Among other consequences, this includes that God has genuine freedom of choice. If an agent[5] has genuine freedom, then there are strict limits to making predictions about that agent's actions. If we could perfectly predict an agent's actions, then there doesn't seem to be any sense in which that agent is 'free to do otherwise'. Thus, since God is truly free, there are strong impediments to predicting God's actions, and further-more the very desire to objectify God as if God were an object and not a subject is itself misconceived.

In short, design arguments have the apparent virtue that they look like standard scientific inferences. But along with this virtue comes a danger—of treating God in the same way that we treat nature (i.e. God's creatures). While these considerations don't show that design arguments are completely theologically inappropriate, they give us some reason to be suspicious.

11.5 The self-undermining problem

The third problem with the fine-tuning argument is that it undermines itself. In particular, the fine-tuning argument notes that the laws (i.e. Einstein's field equations) have the following feature: not many initial conditions would lead to a nice outcome. And since the outcome is nice, it seems that somebody wise needed to intervene to select an appropriate initial condition.

But pause for a moment to ask yourself: what is God's relation to the laws of nature? If God is free in creation, then, as far as we know, God could have chosen a different set of gravitational laws than those dis-covered by Einstein. Let's extend our concept of nice. So far we spoke of a universe as being nice if its conditions allow it to harbour life. Now let us define a usage of nice applied to the patterns called laws of nature. Let's say that laws are nice just in case for almost any initial condition, these laws lead to a nice universe (one in which sentient life could develop). Physicists would describe these nice laws as 'robust' or as 'washing out' those initial conditions that might have threatened to lead to a bad universe.

---

[5] Throughout the chapter, we use the term *agent* to refer to one who has agency; one who acts.

We will now embark on showing the self-undermining problem that arises in design arguments based on fine tuning. In order to do this, we must first 'buy in' to the project of presenting arguments as if God's existence could rationally be questionable. We already explained that this is to misuse the name 'God'. Therefore, in the following we will use a lower case 'g' and present the argument as an argument concerning the existence of some sort of invisible powerful agent called 'god' whose reality is in question, somewhat like the question whether X rays, cathode rays, and N-rays are real (this question was faced by early twentieth-century physicists, and decided by controlled experimentation[6]). The idea is that invoking this agent is meant to help us orient ourselves correctly, or at least better, towards God.

Now, an agent that is able to do all that is logically possible has the power to choose not only the initial conditions, but also the laws. What's more, if we could expect this god to choose nice initial conditions, then we could also expect this god to choose nice laws. What this means is that if the nice initial conditions of our universe confirm god's existence, then the uncongenial laws of our universe disconfirm god's existence.

In order to understand our point, consider an analogy. Suppose that you receive a letter in the mail from an enigmatic billionaire. In the letter, you are invited to visit this billionaire's home on a secluded Greek island, and to participate in a game that he has set up. In this game, you are allowed to choose from one of thousands of gift boxes, with the following condition. You may choose only one box, and you must accept the contents of the box, whether they be good or bad. The billionaire won't tell you in advance what kinds of things are in the boxes, or the odds of choosing various things.

Playing this game doesn't sound like a wise choice. But suppose that nonetheless, you decide to do it. You pack your bags and board the plane to Greece. When you arrive at the billionaire's secluded retreat, he introduces himself to you, and shows you to an enormous warehouse with the thousands upon thousands of small boxes on shelves. He says: you have one hour to choose the box, and you cannot pick up any of the boxes before you make your choice. When you touch a box, then it's automatically the box you've chosen.

---

[6] Cathode rays turned out to be electrons, X rays are a form of electromagnetic wave, and N rays (propounded by René Blondlot) are a delusion.

You realize that there is no strategy to this game—any box is as good as any other. So, you close your eyes, turn a few circles, and walk forward with your arms stretched out. Your right hand touches a box, and you take it down from the shelf. You open the box, and it contains a silver USB thumb drive. You plug the drive into your computer and find that it contains a single file—named 'cure.docx'. You open the document, and the first sentence says, 'The following document describes the cure for cancer.' Incredulous, you give the document to a medical researcher; and quickly thereafter, she confirms that it does describe a cure, and that never again does a human need to suffer or die from cancer.

The box you opened was 'nice'. Does the fact that it was nice provide any evidence about the character of the eccentric billionaire?

Suppose now that after you chose this nice box, you learned that all of the other boxes—thousands upon thousands—contained horrific viruses. If you had opened any one of those boxes, then you would have unleashed a plague that would have caused millions of people to suffer and might have annihilated the human race. All of these other boxes are 'bad'.

This second discovery—that almost all outcomes are bad—provides evidence against the benevolence of the billionaire. And the more boxes there are, the more evidence it provides against his benevolence.

Now, let's return to the case of the fine-tuning argument. If Einstein's theory is true, or at least not misleading, then most universes are bad. So if the style of reasoning adopted in the fine-tuning argument were valid, then it's as if the agent who chose the laws of cosmology was disposed in a hostile fashion towards the emergence of life, but the one who chose the initial conditions was friendly to life. Thus, the evidence works both ways.

An advocate of the fine-tuning argument might object that there are not two agents but one under discussion, so the question of hostility does not arise: the outcome is nice so that is what the agent (god) chose. That outcome was extremely unlikely (goes the argument) under some other assumption, so we have a good line of reasoning—not a proof, but a helpful pointer—to god. However, that argument does not consider the situation in full. The facts, as far as current science can tell us, include both the cosmological model (general relativity) and the nice conditions. We need to ask, does this combination allow us to construct an argument based on probability? The answer is no, because this is not a case of probability but of ignorance. We have not the slightest idea how to assign probabilities to this combination of facts, nor how to say whether they are more or less likely under the assumption of the existence and role of an intelligent creative agent (god) supportive of life. This is because, if one makes that assumption, and if one accepts the line of reasoning (which we do not), then one might suppose that such an agent would be expected to produce nice laws. But we had already supposed that we do not observe nice laws. So what are we to conclude?

For clarity, let us repeat: we do not doubt God's goodness, but our point is about the correctness or otherwise of a certain line of reasoning. Remember that we set a high bar on what constitutes good argument. The fine-tuning argument presents a real conundrum. If this way of responding to evidence is correct, then the two pieces of evidence pull in opposite directions. The niceness of our particular universe suggests that it was created by God (goes the argument). But if this argument is built on a requirement for fine tuning, then it must assume the non-niceness of the laws, so that fine tuning is needed. But then the non-niceness of the laws suggests that these laws were not configured for the promotion of life. So what are we to conclude? Is the universe specially configured to promote life or not? We can't tell. Not from this line of reasoning, in any case. In short, the fine-tuning argument backfires upon itself.

To be clear, we (the authors) don't think that the 'badness' of Einstein's field equations counts against the benevolence of God. This is because we are realistically cautious about our ability to guess correct scientific theories based on our theological understanding. Yes, we recognize that God is both benevolent and wise. But how would God display this benevolence? Already the Old Testament undermines any naïve views of benevolence (e.g. God would make everybody happy all the time). Thus, in our role *qua* scientists, the recognition that God is benevolent has little predictive use, especially not in cases where we're talking about inanimate physical objects or about abstract things such as laws of nature. Already the early modern natural philosophers (such as Robert Boyle) warned strongly against attempts to extract scientific hypotheses from conjectures about God's nature and intentions.

The price of our proper intellectual humility is that we cannot take the existence of our nice universe as confirming that God is as God is said to be in our faith tradition. For the existence of our universe to confirm God's character, we would have had to be able to predict—based on knowledge of God's nature—that God would create a universe like this one. We just don't go in for the game of trying to predict types of universes based on what we only see through a glass dimly.

11.6 Fine tuning of the physical constants

We've been speaking about the fine tuning of the initial conditions of the universe. We pointed out that the initial conditions of the universe wouldn't even have to be fine-tuned if the laws had been nicer. But many versions of the fine-tuning argument don't speak directly about general relativity and about initial conditions of the universe. Some of the most prominent versions of the fine-tuning argument talk about physical constants such as the mass of the proton, or the fine structure constant. These arguments typically run as follows:

> Of all possible values of the constant C, the range of life-permitting values is extremely small. Thus, if naturalism is true, then we shouldn't expect life to be possible. Since life is possible, naturalism is probably false.

This style of argument has some strong similarities to the other version of the argument we presented; and therefore many of our criticisms carry over wholesale. For example, these arguments always assume the structure of the physical laws as a fixed background. But the structure of the physical laws is *not* a fixed background for one who could accomplish anything that is logically possible—God could have chosen a different set of laws, among those that are logically possible, and for all we know, there may be a set of laws for which the life-permitting ranges of constants is much larger.

There is one last gasp at this point for the 'analytical theist' (i.e. the proposer of the design argument). Our criticism of the fine-tuning argument presumes that God has the freedom to choose among different possible laws of nature. But what if God is more constrained than we have supposed? What if in order to accomplish certain aims of His, God could only choose certain sorts of laws, and what if those laws were all not nice in the sense we described earlier? In that case, the existence of bad laws wouldn't count as evidence against God's interest in promoting life.

All we can say at this point is that while this 'just so story' is not incoherent, it nonetheless has little apologetic value—i.e. it has little value as an argument that would move a non-theist to belief. This is because she has little reason to think the story is true. Now, if we knew that science was finished, and that Einstein's theory was correct in all details, then all theists might feel an interest in exploring the question why God chooses to guarantee a good outcome through the means of a balancing act in which the laws of nature are bad. But we (the authors) don't think that science is finished, and we hope that there are better cosmological theories in the future. Thus, we find no reason to seriously entertain this just-so story.

11.7 Is there probably a transcendent creator?

We will now turn to a style of reasoning that is a close cousin of arguments such as the argument from design or from fine tuning. We will present the reasoning and then point out its limitations.

Sometime before his death in 1761, the Reverend Thomas Bayes wrote *An Essay towards solving a Problem in the Doctrine of Chances*.[7] We introduced Bayesian probability in Section 5.1. Proposition 5 gives a description of conditional probability:

> If there be two subsequent events, the probability of the second b/N and the probability of both together P/N, and it being first discovered that the second event has also happened, from hence I guess that the first event has also happened, the probability I am right is P/b.

We will state this more clearly by writing it as a formula. Readers who prefer to skip the formulas can still follow the argument, but we give them for the sake of clarity for those who (like us) enjoy these things. If the first event in Bayes' statement is replaced by a hypothesis h, and the second is taken to be evidence e, and throughout there is some background knowledge k, then Bayes' Theorem states that the probability that we are correct to believe the hypothesis h is

$$P(h|e\&k) = \frac{P(e|h\&k)P/b(h|k)}{P(e|k)}.$$

In this formula the background knowledge k is included for completeness; since it appears in each term it is often regarded as implicit and

[7] Thomas Bayes, 'An Essay towards solving a Problem in the Doctrine of Chances', *Philos. Trans. R. Soc. Lond.* **53** (1763), 370–418.

omitted. The term *Bayesian inference* is often used to distinguish probability that a belief is correct from frequentist probability, which predicts the fraction of outcomes in a series of repeated trials. Putting the numbers from the example in Section 5.1 into Bayes' formula gives the probability that I am correct to believe that I have cancer $0.00001 \times 0.99/0.05 = 0.02\%$. In the box we give another example using coin tosses.

An illustration of Bayesian reasoning

Suppose that the kick-off of a football match is to be decided by coin toss, and a football player wishes to check whether the referee has a fair coin. The referee produces a coin and flips it with the result of 'heads'. He does this twice more, and every time the result is 'heads'. Based on this information, what should the football player conclude?

The answer depends on what the player already knows of the referee. Suppose first that he knows the referee to be a dubious character, who has in his possession a double-headed coin, which he uses as often as not. In other words, we are supposing that the prior knowledge k suggests the two possibilities 'double headed' and 'fair coin' are equally likely. The football player forms the hypothesis $h =$ 'the coin is double-headed'. He has the evidence $e =$ 'it gave a head three times in a row'. The probabilities in this case are

Probability that h is true, given only our prior knowledge k: $P\big(h|k\big)=\dfrac{1}{2}$

Probability that we expect evidence e would happen, based only on our prior knowledge, before doing the experiment:

$$P\big(e|k\big)=\frac{1}{2}\times1+\frac{1}{2}\times\frac{1}{8}=\frac{9}{16}$$

Probability that evidence e would happen if our hypothesis h is right:

$$P\big(e|h\,\&\,k\big)=1$$

Therefore, the probability that our hypothesis is right, given the evidence and our prior knowledge, is (by using Bayes' formula):

$$P\big(h|e\,\&\,k\big)=\frac{1\times\dfrac{1}{2}}{\dfrac{9}{16}}=\frac{8}{9}$$

In this case, then the football player would be justified in suspecting foul play.

But suppose that the football player knows the referee to be not quite so dubious a character. Suppose that nine times out of ten, other things being equal, the referee produces a fair coin. In this case the calculation goes as follows:

$$P(h|k) = \frac{1}{10}$$

$$P(e|k) = \frac{1}{10} \times 1 + \frac{9}{10} \times \frac{1}{8} = \frac{17}{80}$$

$$P(e|h\,\&\,k) = 1$$

Therefore (again, using Bayes' formula):

$$P(h|e\,\&\,k) = \frac{1 \times \dfrac{1}{10}}{\dfrac{17}{80}} = \frac{8}{17}$$

So now the evidence, when combined with the prior knowledge, suggests that the probability of the hypothesis of 'foul play' (the use of a double-headed coin) being true is just under 50%. Perhaps the player should give the referee the benefit of the doubt.

This example illustrates the mathematical method, and it also illustrates an aspect of the human situation. If this experiment were carried out, should the player now change his opinion of the referee, and decide that, instead of a likelihood of 90% for the hypothesis 'referee is fair', the likelihood should now be estimated as $\frac{9}{17}$, which is about 53%? That is, should a mere chance event, which would happen one time in eight even with a perfectly fair coin, cause one person to abandon his confidence in the integrity of another person?

Bayes' essay was published two years after his death with many amendments and additions by a moral philosopher and nonconformist preacher named Richard Price. Price wrote an introduction in which he used the mathematical result to reinforce belief in God, in what might now be called a teleological argument:

The purpose I mean is, to shew what reason we have for believing that
there are in the constitution of things fixt laws according to which things
happen, and that, therefore, the frame of the world must be the effect of
the wisdom and power of an intelligent cause; and thus to confirm the
argument taken from final causes for the existence of the Deity.

Bayes' theorem has subsequently been taken up by the distinguished
Oxford philosopher Richard Swinburne. There are essentially two
components to his argument.[8] First, Bayes' theorem is employed to
make a cumulative case for belief in God, and secondly, one brings in
the idea that, other things being equal, one should prefer the most simple or elegant explanation for a given body of data.

Concerning the first, some apologists use the metaphor of different
strands of evidence wound together into a rope. The strength of the
rope derives from the combined strength of the strands together. The
rope is tolerant of isolated defects in individual strands; fracture theory
explains why the composite nature of a rope gives a robustness against
nicks and scratches which might be fatal to a monolithic structure.
A similar robustness carries over to a case made from different strands
of argument which may, for example, handle different aspects of the
whole or throw various illuminations on complex aspects of the whole.

Rather than this metaphor of strength and toughness derived from
parallel strands, Swinburne builds up a cumulative argument, with
each fresh piece of evidence enabling a Bayesian update on the probability of the correctness of certain types of belief in God. If the prior probability that a belief is true is $P(H \mid k)$, then the Bayesian formula enables
you to update the probability $P(H \mid e\&k)$ that H is true in the light of
fresh evidence e.

To present the chain of reasoning, we must begin with some statement of what assertion or hypothesis we are wishing to investigate. Let
us say the hypothesis is 'the universe has been brought into being by a
transcendent creator who has also made Himself known in human
experience'. We shall call this hypothesis H. The reasoning starts with
the observation that there is a universe. By what is generally known as
the cosmological argument this gives some probability that H is so. The
universal operation of natural laws provides fresh evidence that enables
you to update the probability of H. If the probability that God would

[8] Richard Swinburne, *The Existence of God* (Oxford University Press, 1979; new edn
2004).

create a universe with consistent natural laws is greater than the probability (whether or not God created the universe) of consistent natural laws occurring anyway, then the probability of H being true increases. The change is given by the ratio $P(e \mid H\&k)/P(e \mid k)$.

This process of updating the probability of H can be repeated for each new piece of evidence. Pieces of evidence might come from the existence of human and animal organisms, the observation that humans (and maybe animals) are conscious beings with sensations, beliefs, thoughts, desires, and purposes. There is evidence from history, and the existence of documents which describe experience of God, including miracles associated with the life of Jesus. There is the evidence for the resurrection of Jesus, or if you prefer, evidence for contemporary confidence in the resurrection of Jesus. Finally, there is the cumulative evidence of religious experience. For each of these, if $P(e|H\&k) > P(e|k)$, then the successive probability that H is true increases. One must also include counter-evidence of course. This is the set of cases where $P(e|H\&k) <$ $P(e|k)$. The probabilities are hard to assess. One may incline to the view that natural disaster, human tragedy, and the freedom allowed to human evil all conspire to produce low values of $P(e|H\&k)$. Conversely, one may advance reasons why these are not unexpected under the hypothesis H.

Next, we bring in the general scientific principle that given a choice of explanations which fit the facts, the simpler explanation is to be preferred. A favourite example of Swinburne is to suppose that a burglary has occurred. In the subsequent police investigation, various clues come to light, each of which individually is amenable to more than one explanation. Common to all the candidate explanations is that the suspect John committed the crime. Given the choice between the multiplicity of separate explanations required if John is innocent, and the single explanation that John is guilty, the simpler account is preferred. Similarly, in science the simplest theory that fits the facts is to be preferred over more complicated theories.

Now let us critique what we've just discussed.

Scientists who do not believe in God may not be persuaded to belief by Swinburne's use of Bayesian reasoning. But even those who warm to his conclusion may find the steps hard to subscribe to, for reasons which can be divided among the two components identified earlier.

First, when scientists see a formula, they instinctively want to put values in. Bayes' formula has essentially three quantities on the right-hand side, to enable one value to be calculated for the updated probability that H is true. So the scientist will wish to know what is the probability that God would create a universe; what is the a priori probability of a universe existing; and what is the a priori probability that God is approximately as the hypothesis H suggests. The difficulty is that we have simply no idea what values to attribute to each of these. A value between 0 and 1 is needed, but we have no objective criteria for assigning one. The same goes for many of the subsequent steps in the Bayesian chain of reasoning. Thus, while a scientist may warm to the concept of growing confidence with successive pieces of evidence, it seems to be impossible to put quantities into the Bayesian formula to achieve any reliable conclusion.

Second, the notion that a simple explanation is to be preferred to a complex one is not false, but this issue has some subtlety.

The issue in scientific work is first and foremost accuracy, rather than simplicity. That is, a simpler model of some phenomenon will not be preferred if it also makes inaccurate statements about some other phenomena. In order to assess this, one may need to explore implications of the given model that are outside the context in which it was first proposed, and this is not always easy to do. Also, there is the issue of experimental uncertainty. Data are always imprecise at some level, and it can happen that one cannot easily tell whether a simpler model is lacking in significant precision which a more complicated model can furnish.

There can also be differences of opinion on what is the more simple or elegant or cogent set of ideas.

Consider, for example, the use of complex numbers in science. The technical term 'complex number' refers to the use of numbers which, when multiplied by themselves, may possibly give a negative result. The basic example is the number (often referred to by the letter i) which, when multiplied by itself, gives minus one. In standard notation, $i = \sqrt{-1}$.

This is not a number in the everyday sense of the word, but it can be shown that mathematical rules of addition, subtraction, multiplication, and division still make sense when working with i, and indeed, many mathematical problems can be greatly simplified that way. Now, when we approach physical science such as physics, chemistry, and engineering, it is always the case that all the required mathematics can be written down without ever introducing $\sqrt{-1}$. This is because a single equation involving complex numbers can always be written as two equations involving only real numbers. However, doing science that way is like fighting with one hand tied behind your back. It is much easier to let $i = \sqrt{-1}$ come in, and thus simplify the equations. In this choice, one gets equations that are simpler in one sense and less simple in another. They are simpler in that fewer symbols and operations are involved, but less simple in that the mathematical quantities being manipulated (the complex numbers) are themselves richer and less straightforward than real numbers.

This example of complex numbers illustrates a recurring theme in science. One could multiply examples. We will show two more.

When learning about electric and magnetic fields, first at school and then more fully at university, students are almost always introduced to the ideas by using the language of *vectors*. The electromagnetic field at any given place is described using a pair of vectors. A vector is like an arrow, and it can also be thought of as a set of three numbers written down in a row or a column. Later, however, when we want to go further, especially when considering motion at high speeds, we teach the students to regard these vectors as two parts of a single mathematical entity called a tensor. A tensor cannot easily be pictured, but mathematically it can be thought of as like a matrix or a table of numbers. The transition here is like the transition from real to complex numbers: the use of the richer mathematical tool (the tensor) allows the equations to be simpler, but it is hard to argue that the tensor is itself conceptually simpler than the pair of vectors, because if it were, then we would set out by teaching that method in high school!

For our third example we will make a jump to the life sciences. Suppose one wants to consider the population statistics of rabbits, for example. A basic element in the discussion might be 'one rabbit'. But a rabbit is a hugely complicated thing! It is much more complicated than each of the individual molecules or smaller particles that form its body. The simplicity of being able to say 'one rabbit' is bought at the expense of working with a highly complex basic unit. But this is genuine simplicity and correct insight, nonetheless, because a discussion purely and directly in terms of the movements of all the individual molecules would not be a simpler nor a simplifying discussion.

Now let us return to the application of the notion of simplicity to thinking about God. The central issue here is that different people have vastly differing opinions about whether an appeal to God can correctly be called simplifying. Some people judge that what is claimed of God, such as that God can attend to the prayers of billions of people, must imply that God is fantastically complicated and therefore no explanation invoking God can be simpler than one that does not. Other people judge that the simplicity associated with God is more like the simplicity associated with deeper explanation, like the move from real to complex numbers or from vectors to tensors. We (the authors) have some sympathy with both these reactions, so we are not going to be able to offer a neat resolution. We have already expressed our reservations about the whole project of trying to place God in the role of an intellectual tool for humans, even including the role of 'best explanation'. We think God can be 'best' in the sense of 'loveliest' or 'affirmer of attitudes we should learn' but not in the sense of 'useful to filling in gaps in a sequence' or 'confirming our prejudices'.

This brings us to two further observations that may be relevant in an appraisal of arguing from Bayesian probability.

First, when we talk about God we use concepts from our material experience. That is the best we can ever do. But we should not confuse concepts that are to a greater or lesser extent metaphorical with the actual reality. To talk of God as a hypothesis with a probability of being true reveals a woefully incomplete appreciation of who God is. It may even be fundamentally misconceived, like talking of a person as if they were an object. It may not even be meaningful to think of God as a thing whose existence can be predicated, because it would be like thinking of a circle whose circumference was rectangular. That is to say, it would be a self-contradictory exercise. To be clear, the self-contradiction here is not the attitude of turning to God with a desire to learn, but the attitude that supposes God could possibly be our property or invention or deduction.

If God is involved in this world, then that is why we can talk about God at all, but if that One is at the same time utterly different from our categories of discourse about everyday objects and concepts, then we must watch our language.

Second, the context in which a question is asked makes a lot of difference to how useful the question is. Every experienced scientist knows, for example, that the timing of questions can make all the difference in a professional career. Ask a question too early, and the techniques may not be available to address it. Ask a question too late, and the rest of the scientific world may have moved on. So too for an individual asking questions about the existence of God. The person who does not know God may find it helpful to think about the evidence that God is there. If such a person finds the Bayesian approach useful, well and good, though it may not be the best approach and it is certainly not the only one. The person who already knows God may rather quickly lose interest in discussing God's existence.

Consider an idealized case of courtship. Two people, say Bob and Alice, meet and find each other interesting. At first, neither of them has nearly enough information to make a reliable judgment about whether the other would make a suitable spouse. So what should they do? If Alice were to follow W. K. Clifford's maxim, i.e. of believing in proportion to the evidence,[9] then she would give little credence to the idea that Bob would make a good spouse for her. And yet, she might decide that she wants to learn more about Bob—in hope of discovering that he has the qualities that she seeks in a spouse.

When it comes to knowledge of other persons, our epistemic duties are different, in important ways, from our epistemic duties in science. In science, there is no question of trusting a hypothesis, or of giving it the benefit of the doubt, or of loving it. In contrast, when it comes to other persons, the 'scientific outlook' can be completely inappropriate. Living often requires us to act in ways that go beyond the available evidence. A person may also choose, or not, a course of action that is likely to yield more or better understanding of evidence. One could think of this as a Bayesian process, albeit one that may be as much emotional as rational, with potential for both joy and heartbreak. In the case of personal relationships, it is rational to 'get to know a person'. We argue that the same is true in the case of religious belief. That is, it is rational to explore a relationship with God, in the hope that God is everything that others have described Him as being.

Suppose now that Alice and Bob are a married couple. Should Alice ruminate on whether 'Bob is a good spouse' or 'Bob is worthy of my love', or on what methodology to use to determine whether she is justified in believing these claims? In a strong marriage Alice would not focus on the evidence for or against such questions—the time for that is over. In a healthy marriage, she would focus her efforts on cultivating the relationship. For the person who knows God the question of God's existence

---

[9] 'It is wrong always, everywhere, and for anyone to believe anything on insufficient evidence.' W. K. Clifford, 'The Ethics of Belief', *Contemporary Review* (1877); reprinted in *The Ethics of Belief and Other Essays*, ed. T. Madigan (Prometheus, 1999), 70–96.

no longer helps, or even seems appropriate. God is not an object to be examined and evaluated. For such a person there are other more pressing and fruitful questions, such as how does God want me to live, and what resources are available for that purpose?

Earlier we used the word 'loveliest'. It is helpful to note that this does not need to suggest the notion of a powerful king, the way God has often been portrayed in works of human imagination. Such language is usually a way of getting at the notion of justice. When people have been subject to unjust authority, they have wanted to say that God is both fairer and more powerful than whatever the temporal authority is, and they have expressed this using the metaphors available to them. In consequence, however, it was all too easy for Christianity to find itself caught up with notions of royal rule and imperialism. However, modern-day followers of Jesus are becoming more aware of this chequered history and more interested in asking what justice really looks like. More and more we are finding that Jesus' demonstration of 'existence for others' is itself enough; it really is through love and not the projection of power that anything worthwhile gets done, and justice will consist in all things being seen to be exactly what they are. With this in mind, we are willing to let go of all pictures and simply affirm, with Jesus, that God is about the gut-truths of human life: that hearts that can mourn are healthy hearts, and that people who don't dwell on insults are the ones who show what life is; that it is the rich not the poor who are more often the blind or exploitative, that all people are valued by God in the same way that a good parent values their children equally, and finally that we can cease judging each other in a superior way because God guarantees that all things will finally be seen to be exactly what they are.

11.8 What's wrong with God of the gaps?

There is another, more general, objection to all design arguments—the objection that these are 'God of the gaps' arguments. Such arguments run like this: feature X of the universe has no known naturalistic explanation. But everything has an explanation, so the explanation for X must be God.[10]

---

[10] Some more sophisticated design arguments claim that feature X could not *possibly* have a naturalistic explanation. From our point of view, such arguments are presumptuous about the future course of natural science.

There has been a lot of back and forth about the merits and demerits of this kind of argument. On the one side, we have contemporary advocates who think that these sorts of arguments have failed, and that we should learn from our past failures. On the other side, there are a few Christian thinkers who say that it would be presumptuous to rule out in advance the fact that there might be some events that could only be explained by God.[11]

We are inclined strongly against gap arguments, but for quite different reasons than those of the typical naturalistic critic. We think that gap arguments are bad because there are certain rules of good thinking—rules that were written on our hearts by our Creator. But note that our critique of gap arguments suggests another way of perceiving and responding to God: good reasoning finds its root and support in the character of God; hence, it is God's being that ultimately explains why we are able to perceive that gap arguments are bad!

11.9 Leibniz's criticism of Newton

Let's consider a historical episode, where one scientist criticized another for shirking his scientific duty.

When Newton proposed his law of universal gravitation, the question arose as to why the planets orbit the Sun in roughly the same plane, and in the same direction. Of course, Newton's laws can tell us why a planet follows a certain trajectory, given certain initial conditions. But Newton's laws have nothing to say about why these initial conditions obtained in the first place.

Newton's response to this problem was to invoke God's providential arrangement of nature. According to Newton, the reason that these unlikely initial conditions obtained was a free action of the Wise Author of the universe. But Newton's answer here was rejected by none other than Leibniz, his arch-nemesis in all matters scientific:

> For to have recourse to the decision of the author of nature is not sufficiently philosophical when there is a way of assigning proximate causes.[12]

You might be tempted to think that Leibniz is further down the path to secularization than Newton, but that would be a misconception. For

---

[11] We offer a detailed critique of so-called Intelligent Design in Section 12.5.

[12] G. W. Leibniz, *Philosophical Essays*, ed. R. Ariew and D. Garber (Hackett, 1989), 331.

Leibniz, God is at the centre of everything. Indeed, for Leibniz, God is the paradigmatic craftsman, and such a craftsman would build things to realize his ends, without his ever having to step in to make adjustments. For Leibniz, one who trusts in God should expect to find no gaps in nature.

History has validated Leibniz's instinct to search for naturalistic causes: contemporary astrophysics explains the coplanarity of the orbits of planets in our solar system. To do so, we only need to refer back to earlier initial conditions in the early universe.

But doesn't Newton get the last laugh here? Leibniz was right about the particular case. But Leibniz's victory was temporary, because now there is another thing that science doesn't explain: the special initial conditions of our universe. If Newton were alive, wouldn't he be right to insist that at every stage of science, there will be some initial condition that cannot be explained by science?

This last statement hides an ambiguity. On one interpretation, it says that there is some particular initial condition that will never be explained by science. On another interpretation, it says that at each stage of science, something will not be explained by science. The first interpretation is strong and implausible, because science always seems to find a way to explain things that seem puzzling or surprising. The second interpretation is more reasonable.[13] We can grant that for any particular contingent event that humans have observed, science has eventually been able to give an account thereof. But it's just as clear that at no time in history has science explained all the circumstances of any event. Thus, the most reasonable prediction is that in the future, science will march forward explaining more and more...and that there will always be more things that science does not yet explain.

This picture of science—as infinitely deep—fits with the history of science and presents us with a beautiful picture of where science might go in the future. This picture also happens to sit most comfortably with a theistic outlook. For if the universe has a transcendent source, then the call to understand the universe will never be satisfied by finite beings. At each stage in the history of science, we are called to understand more and better. In this picture, there can be no such thing as a 'final theory'.

---

[13] Martin Rees, 'Is There a Limit to Scientific Understanding?', *The Atlantic* (6 December 2017); available at https://www.theatlantic.com/science/archive/2017/12/limits-of-science/547649/

We should be clear at this point to contrast our view with a certain sort of rationalistic theism that suggests that God 'had to do things a certain way'. For example, some theists argue that God of the gaps arguments are bad because they are an affront to God's wisdom—that God would never create a world with gaps in it.[14] But far be it from us to say what kind of world God would create! No, we are not confident in our ability to predict what kind of world God would create. What we do know are the commands that God has written on our hearts—and one of these commands is to seek a deeper understanding. Thus, if we see a gap in the scientific account, we don't presume that God wouldn't create a world like this. Rather, we assume that our scientific task is probably not yet finished.

How then should we apply Leibniz' lesson to contemporary physical cosmology? There is one potential disanalogy: Newtonian physics wasn't explicitly concerned with the origin of the universe in the way that contemporary physical cosmology is. And don't we have in principle theological reasons for ceasing the search for naturalistic explanations when we're talking about the beginning of the universe?

Let's not forget, however, that the Big Bang is not necessarily the creation event that is spoken of in the first verses of Genesis. When it comes to building an integrated view of reality, religious believers can do no better than to equate their theological creation story with the best current scientific account. But this synthesis must always be carried out with extreme caution, for the scientific story *will* develop and change over time. If we tie the theological account too tightly to the scientific account, then the next scientific revolution might be mistaken for a crisis of religious belief. A properly framed religious conviction does not take the form, 'this is what happened, and therefore it must be owing to God,' but rather, 'whatever it was that happened, the fact that it can embody goodness is owing to God.'

The example of Leibniz shows why theists ought to be troubled by gap arguments. But you might wonder then—how does our view differ from an atheistic view? The short answer is *fundamentally*. The longer answer is that our motivation for eschewing gap arguments comes directly from our theistic commitment, and from our theistically oriented view of science.

---

[14] This picture is sometimes called the 'functional integrity view', and it has been championed by Howard J. van Till and Diogenes Allen, among others. See also M. J. Murray, 'Natural Providence (or Design Trouble)', *Faith and Philosophy* **20** (2003), 307–27.

11.10 On the retreat?

The naturalistic opponent of theism might try to interpret our argument in this chapter to mean: 'Here go the theists again—as soon as science advances, they retreat.' This is one of the New Atheists' favourite arguments: that theism is constantly losing ground in a battle against science. But we (the authors of this book) were never inclined to place our hopes in this sort of design argument—the sort that says some particular feature of nature can only be explained by divine action. For this sort of design argument comes into tension both with our theological inclinations, and with our (theologically informed) scientific inclinations. First, our theological views hold that God is the unique creator of nature and a participant in a covenant with us, and so is not an appropriate subject of scientific scrutiny. Second, our theological views suggest that human beings are called to seek an ever-deeper understanding of nature—to strive to make puzzling phenomena (such as fine tuning) intelligible within a better scientific framework.

But before we make our positive case, we need to address a common argument. The argument goes something like this:

> In the past, whenever people have ascribed some 'wonderful phenomenon' to Divine action, it has later come to pass that scientists have found a purely naturalistic explanation for that phenomenon. By inductive reasoning, we should expect that in future cases, there will be naturalistic explanations for any phenomena. Thus, there is no need to posit God as an explanation for anything in our world.

The intuitive force of this argument masks some serious weaknesses. For example, this argument has precisely the same structure of one sort of argument against the rationality of science! The argument we are thinking of runs as follows:[15]

> In the past, the best scientific theories have proven to be false. (For example, the phlogiston theory of combustion is false, Newtonian gravitational theory is false, etc.) Thus, we have every reason to expect that the best theories of current science are also false.

There are myriad ways to rebut this argument. One way is to reject the claim that past theories have been simply false, but to say instead that

---

[15] This argument has been named the 'pessimistic meta-induction' by philosophers of science. It's called a meta-induction because it's an inductive argument *about* science. See https://plato.stanford.edu/entries/scientific-realism/#PessIndu

they were approximately true—and that our current theories are better approximations to the truth. But then, in just the same way, it is reasonable to say that past attempts to see God's hand in nature were approximately true, and that current attempts to see God's hand in nature are better approximations to the truth. Indeed, we are quite happy with this way of assessing the situation. Consider, for example, biological evolution. Our ancestors might have seen the structure of the human body as a sign of a creative intelligence. We now know that the human body is the product of millions of years of biological evolution—and so we are no longer tempted to think that somehow God directly and immediately created the human species. Nonetheless, our sense of wonder is only strengthened by coming to see the biological explanation for the human body. Our wonder at the result is not *displaced* by a more complete understanding of the physical process by which it was produced, but rather *augmented*. Francis Collins is an American physician-geneticist noted for his discoveries of disease genes and his leadership of the Human Genome Project; he is now director of the National Institutes of Health (NIH), the US government funding agency for medical sciences. He has been open about his Christian faith.[16] Like him, we stand in awe at the intricacy of the genome, and the sorts of physical processes that worked together in its evolution. Remember, we are not trying to construct a proof of the existence of some hypothetical entity. We are simply appreciating the world as it is and reacting with our whole heart and mind.

We really mean it. We not only don't mind being part of the natural world and an outcome of its processes, we *like* being a part of the natural world and an outcome of its processes. We are grateful that this very world, the world as it is, has furnished our opportunity to be and to have some inkling of God and to participate in God's creativity and grief and joy.

The progress of science has closed many gaps. One might be tempted to infer from this that it would be unjustified to ascribe any natural phenomenon to God—because future science will most likely close the gaps in the current scientific story. We have argued that this sort of thinking is a misunderstanding of the situation. To see God's hand in nature doesn't mean believing that God is an additional causal variable

---

[16] Francis S. Collins, *The Language of God: A Scientist Presents Evidence for Belief* (Simon and Schuster, 2006).

that is required for an adequate scientific account. On the contrary, belief that God was involved in some event suggests that there is some intelligible, interesting, and beautiful natural mechanism. God's involvement is first to make it possible for the event to be; next to know and affirm that the event is precisely whatever it is (herein lies the hope of justice); next to participate in a way which respects the integrity of all parties, and works for whatever good can be promoted through that particular nexus of interactions.

We (the authors) feel that our primary duty is discharged if we've removed some obstacles to faith, especially those obstacles that rest on bad thinking. For it's not within our power to work a change within a person's heart—and that's what it takes for somebody to follow Christ. But we're also not going to leave you without some advice and guidance. We think that there are plenty of reasons that tip the scales in favour of theism. We will mention a few examples from what would be a long list. First, our belief that the universe is God's creation provides a compelling rationale for doing science—i.e. a reason to 'keep seeking' for *logos* in the universe. Second, theism is liberating: it endorses our natural instincts towards thankfulness and hope, and it helps people to help one another. Third, an atheist account for the origins of the Christian movement seems to involve too much special pleading. Fourthly (and we will finish here with this, though there is much more one could say), it's an empirical fact that humans, when confronted by the wonderful in nature, feel moved beyond nature and toward the transcendent. There must be some reason why we have this impulse, and we aren't convinced that it's merely an evolutionary spandrel. Rather, we believe that our impulse toward the transcendent is good; and hence that it's rational for one to turn to God when one encounters the wonderful in nature.

11.11 Why do science?

Why do humans desire to understand the world? Please think about that question for a moment before we continue our discussion.

'Why' questions can mean different things in different contexts. Consider, for example, a question that my wife might pose to me: 'Why do you love me?' Suppose that I answered her by giving a scientific explanation of the following sort: 'I'm a member of the human species, which has evolved in the following way...Since humans engage in sexual activity in order to reproduce, etc.' She is not likely to be pleased with my answer. Even worse, imagine that I state my answer in the language of fundamental physics—e.g. I describe myself as an excitation in a quantum field evolving according to, say, the Dirac equation. If I give a really good physical explanation, will it become obvious that I had to love her, because I didn't have any choice about the matter?

But if you're a well-adjusted human, then you know that I missed the point of my wife's question. She doesn't want a *scientific* answer to my question, she wants a *human* answer. The kind of answer that would meet the situation truthfully might be something like: 'I just do, and I feel it is not about any particular ability you may have; it is about who you are. But since you ask, I love that you always have wise things to say about any topic. Plus, you have great eyes!'

The case of my wife's question is clear. But it's not just in interpersonal relationships where 'why' questions have different sorts of answers. I'm thinking specifically of cases where a 'why' question calls for an answer in terms of *reasons*. For example, suppose that somebody asks me:, 'Why do you believe that $1 + 1 = 2$?' There are two different things that this person might be asking. On the one hand, this person might be asking for a *scientific explanation*, in which case I should cite some initial conditions and laws of nature that eventuated in my having this belief. But that's not typically what a person wants to know when they ask why you believe something. What they want to know is your process of inference, or your logical *reasons*. So, in the current example, if somebody asks me why I believe that $1 + 1 = 2$, the best answer I could give them would be to give a proof of that fact (perhaps from the axioms of Bertrand Russell's *Principia Mathematica*).

Why then do humans desire to understand? As we've seen, we can take this question in two different ways. On the one hand, we can take this question to be asking for a scientific description of the processes that led to our species having certain dispositions. On the other hand, we can take this question to be asking: 'For what reason do human beings seek scientific understanding?' There are various similar questions in the neighbourhood, such as 'Why do we value scientific understanding?'

Let's take it as agreed that, other things being equal, the impulse to do science is good—one that should be encouraged and fostered. It's on this basis, i.e. our sense of the intrinsic goodness of science, that we are averse to gap arguments. But now we have a question for the naturalist critic of gap arguments: why do you think that gap arguments are bad, and why do you think that scientific understanding is good? And let us remind you, we are not asking for a description of why humans are disposed to think that scientific understanding is good. We are asking why it really is good. We are asking for a story about what makes it good.

On this point, natural science can only be silent. Natural science is well suited to explaining why humans have certain inclinations. But natural science is simply not capable of addressing the question of whether inclinations are good or bad. To borrow a phrase from Nietzsche, natural science is beyond good and evil. (Unsurprisingly, Nietzsche thought that scientific understanding is *not* intrinsically good. He thought that the belief that it is inherently good is an inheritance of the Christian tradition, and one that should be rejected along with Christianity.)

What then can a metaphysical naturalist say about the value of science? One possible response is that scientific understanding is of instrumental value. For example, one might take a utilitarian view according to which all value is grounded in pleasure, in which case scientific understanding would be valuable for the various pleasures it provides. But we eschew all attempts to reduce the value of science to its ability to fulfil other human desires, such as the desire to experience eureka moments, or the desire for new technologies, or the desire for a growing economy. What account, then, can we give of ourselves that would affirm the intrinsic value of science?

We think that the most satisfying answer to this question is that understanding is an intrinsic good, and the capacity to understand has been woven into the fabric of the universe in such a way that it finds expression in creatures such as humans. In short, humans were created to understand—each other, and the physical reality in which we live and breathe. Note that this answer is not intended to deliver a scientific explanation of how we came to have the impulse to understand. For one, such a scientific explanation can never tell us that we *ought* to have this impulse—it can only tell us that we do tend to have this impulse. The theistic explanation is really a moral or teleological explanation, as it refers to the intentions that God had in creating us. It does not tell us how God created us, and how God implanted within us the scientific impulse. In fact, it seems that God allowed evolution to run its course, and that evolution provides an adequate explanation for why we do, in fact, have this disposition in a physiological sense. But we weren't asking the question of what mechanisms led us to have this disposition. We are asking: what's the point of our having this disposition?

To see this more clearly, consider the question: 'Why do you value Mozart's music?' There are really two kinds of question here, and accordingly two kinds of answers. On the one hand, you might answer by explaining how evolution favoured genes that manifest in the experience of pleasure when listening to such music. But that's not the kind of answer that is typically being sought. If a person asks why you value a certain kind of music, they are typically looking for *reasons* rather than for causal antecedents. When you answer by giving reasons, you aren't just saying that something *is* the case, you are also attempting to justify your attitude.

In contrast, consider the question: 'Why did Adolf Hitler cause so many people to be murdered?' You might adduce some causal antecedents, e.g. socioeconomic circumstances, Hitler's childhood, perhaps mental illness, etc. You would not, however, cite Hitler's own *reasons*. Indeed, it would be morally callous to do so unless you immediately went on to say that you didn't consider that they are good reasons.

Now, the question 'Why is scientific understanding valuable?' could similarly be understood in two ways: as asking for causal antecedents, or as asking for reasons. The question might mean 'Why do humans feel this way?' or it might mean 'Why ought I feel this way?' For the former question, it might be appropriate to cite physical processes and mechanisms that led to this situation. However, since 'is' doesn't imply 'ought', citing physical mechanisms and processes cannot answer the question, 'Why ought I feel this way?' This second question asks for *reasons*, and we believe that the best reasons have to do with persons and relationships: scientific understanding is valuable because it is part of our loving response to God, each other, and the wider world. It is a movement of the heart, and an outcome of the fact that God created the universe with the aim of producing beings who can love as He loves.

It is entirely consistent with this to note that scientific understanding is a profound source of human satisfaction and pleasure, and—dare we say it—fun!

11.12 Presuppositions of science

We have steadfastly rejected claims to the effect that science can lead us to God. We think that approach is backwards, for belief in God is more fundamental to the practice of science. One way of putting this point is to say that belief in God is a *presupposition* of scientific practice. To say that a belief is a presupposition of a practice means that you must implicitly believe in that thing in order to be able to carry out that practice. But is it at all plausible that somebody must implicitly trust in God's existence in order to do science? We think that the answer to this question is Yes, but in a nuanced sense.

As a preliminary to defending our answer, let's consider a more general question of whether the practice of science has any presuppositions whatsoever. We think that the answer to this question is a definite Yes, and we draw our support here from a famous episode in the history of modern philosophy. The Scottish philosopher David Hume pointed out that there is a vicious circularity involved in any attempt to justify claims through inductive reasoning. Recall that a universal claim, such as 'all ravens are black', seems to be confirmed by observing particular instances, such as 'this raven is black'. The question then arises: how many instances of individual black ravens must we observe in order to be justified in believing that all ravens are black. For our purposes, it's enough to assume that the more instances we observe, the more we are justified in believing the general claim.

But why, asked Hume, does observing more instances supply more justification for believing the general claim? It's always possible that the instances we've observed were atypical. There is never a guarantee that the next instance we observe will be the same. So, it's not pure logic that connects the instance, 'this raven is black', to the general claim, 'all ravens are black'.

Hume then points out that we could attempt to shore up the argument by citing some general principle such as the uniformity of nature: *the future will resemble the past.* But now why should we believe that general principle? That general principle is not a truth of logic, and so the only way we could believe it rationally is if we could justify it by an inductive argument. In other words, we need to use an inductive argument in order to justify belief in the uniformity of nature; and we need to be able to justify belief in the uniformity of nature in order to be justified in using inductive inference.

Obviously, we now fall into either a circle or an infinite regress. In both cases, we run up against a failure to justify one of the fundamental principles of scientific reasoning.

Hume's response to the 'problem of induction' was a combination of lapsing into scepticism and falling back on raw animal impulses. Whether or not humans are justified in using the rule of induction, Hume said, they just can't help themselves—they will continue using it. Thus, for Hume, there is no sense in which science is *rational*; it's just something that human animals do.

Now, along came the German philosopher Immanuel Kant. And Kant was not satisfied with Hume's scepticism. In reply to Hume, Kant said: scientists know things, and so induction must be justified, and so nature must be uniform. In short, Kant turned things on their head. This famous reversal might be considered cheating, but nobody has offered a better story about the normative force of scientific reasoning, i.e. about what makes scientific reasoning sound or *good*. In short, science proceeds on the supposition that nature is uniform. This supposition is not exactly 'blind faith', because blind faith doesn't have reasons of any sort. The belief in the uniformity of nature is justified in a backward sense for what it enables us to do—it enables us to do science.

If Kant is right, then science operates with certain presuppositions about the uniformity and intelligibility of nature. Now, the belief that 'nature is intelligible' is strikingly similar to belief that intelligence is a precondition of nature—that nature exists only because intelligence first existed. When we investigate the natural world, we don't assume it to be like random scribblings on a scrap of paper. We adopt the attitude that it is intelligible, like a novel or a treatise. In fact, we can rephrase the statement 'nature is intelligible' as saying, 'it is *as if* nature were the result of an intelligent act.' Thus, the investigation of nature presupposes that the world is *as if* its most basic qualities have been instilled by intelligent action.

11.13 From wonder to belief

The sense of wonder that one feels in the face of nature—for some people, it suggests the transcendent, and for some people it motivates a search for scientific intelligibility. We think that in a well-adjusted human life, these two attitudes will coexist, and they will complement and mutually reinforce each other. But as for the first attitude, the feeling of standing before the transcendent, is it merely a subjective 'warm and fuzzy feeling', and so not subject to rational scrutiny? Or if it is subject to rational scrutiny, then what are its credentials? Is it reasonable to become more convinced that our true origin and widest context *comprehends* who we are, or should a person steadfastly resist that impulse?

Consider one of the examples from the beginning of this chapter. When I (HH) first looked at my newborn son, I was filled with a sense of wonder, and a sense of thankfulness. I was inclined to think, 'Only God could create a being like this.'

Was I being rational? Of course, I wasn't being fully rational! I rarely am. But the question is, was I being less rational than I am at other moments, for example, when solving a mathematical problem? I'm confident that I wasn't being irrational. In fact, my response seems to be utterly appropriate to the situation, and even morally good. An irrational response is a *blameworthy* response, and my response was not blameworthy.

Nonetheless, I am *not* suggesting that the following sort of argument is valid:

> The existence of this child is so wonderful and amazing that only God could have brought it about. Therefore, God exists.

It's clear to me that this argument is bad. It does not nearly reach our high bar. One problem with the argument is that there is a perfectly good naturalistic explanation for the existence of my son. So, in one sense it's just false to say that 'nothing but God' could have brought about his existence.

Let's consider the situation more carefully. I'm looking at my newborn son, and my heart leaps—not just toward the boy, but higher—toward a less articulated thought that there must be something more to life than passing on genes from one generation to the next. But my response isn't just a 'warm fuzzy' feeling in my heart. Rather, my response amounts to a sort of increase in my conviction that the universe is intelligible—that there is intelligence behind it. So, it's not just an emotional change, it's a cognitive change. My inclination to trust in a personal God has been strengthened. But could this cognitive change be rational despite my lack of any clear inference of God from the phenomena?

The answer is Yes, this change of opinion can be rational. A person doesn't need to have anything resembling a scientific abduction in order for his or her shifts of opinion to be rational. If you disagree with that claim, then I wonder if you live consistently with your principles, and I wonder if you'd be happy if other people around you lived by your principles. It seems to me that consistent application of those principles would cripple the normal emotional and intellectual dynamics of a human life. For example, I (HH) tried to choose a career (being a professor) that suited me. I knew that I wasn't suited to be a professional athlete, nor was I suited to being a medical doctor. But my confidence in the claim 'being a professor is a suitable career for me' depends in an intricate way on my life experiences. For example, if I feel that I've given a good lecture, then my conviction in this claim increases. On the other hand, often when I'm sitting in a department meeting, my conviction in this claim decreases. But it would be ridiculous to think of this claim as standing in the same relation to my life experiences as a typical scientific hypothesis stands in relation to data collected in the laboratory. One basic disanalogy is that I'm not directly interested in testing this claim, at least that is not my aim most days, and so I am not attempting to find phenomena that supports or undermines it. Instead, it's a question of whether adopting the attitude 'this career suits me' is appropriate to my situation.

In general, it's not irrational for me to change my confidence in this claim depending on how my life is going. Certainly, some radical changes of attitude would not be suitable. For example, if I gave a bad lecture, and I concluded on that basis alone that I was definitely not cut out to be a professor—well, that would be rash (as my wife likes to point out). Similarly, it might be irrational if I continued to hold on to the belief that I ought to be a professor even if my teaching evaluations were universally poor, I couldn't publish any papers, and I found the work to be dull, etc. Thus, while the claim about my suitability for this career is not a scientific hypothesis, it is open to rational re-evaluation in the light of fresh evidence. This would be another example of Bayesian reasoning.

Changing one's attitude toward transcendent questions, e.g. whether or not there is compassion and intention, not just abstract beauty, in the transcendent, is similar in relevant ways to changing attitudes about one's career choices. A human life is filled with various experiences, and these experiences push us in subtle ways toward, or away from, life choices. It would be rational for me to increase my confidence in my career choice if I received some prestigious award, or even if I just found myself enjoying the job. Similarly, it would be rational for me to increase my confidence that the universe was created by a Person when I experience the profound satisfaction that accompanies scientific understanding.

11.14 Conclusion

We've been harsh on the cosmological fine-tuning argument—but not because we reject its conclusion. Indeed, we feel a need to be harsh on the fine-tuning argument *because* we accept that God created the universe. For such acceptance brings along with it a call to try to understand nature. Furthermore, if God did, in fact, create the universe, then the demand for scientific understanding is genuinely *infinite,* and will never be satisfied in our lifetimes. Thus, for any natural phenomenon, no matter how puzzling or wonderful, we are called to seek a deeper scientific understanding.

Another reason we find the fine-tuning argument troubling is because the statement 'God created the universe' is different in kind from any hypothesis of empirical science. 'God' refers to an uncreated being, or to uncreated being *tout court.* Such being is *not* an appropriate object for *natural* science, nor even accessible to encounter by a method of purely passive scrutiny. Furthermore, the claim that God created the universe doesn't tell us anything about *how* such an outcome could happen, nor about how the universe functions. There is an empirical paucity to 'God created the universe' precisely because this statement is different in kind to empirical hypothesis. This is not a weakness of the statement, but simply an example of the fact that, in order to be true, any statement must be true to its own nature, to what sort of statement it is. In a loose comparison, this statement is akin to statements such as 'these sounds are beautiful' or 'we should treasure this bowl because it was a gift from my late husband'. The fact that the bowl was a gift is not indicated empirically by any property of the bowl, but nevertheless the inclination to make the bowl the focus of an added care is not an empty or valueless inclination. In short, to say of a statement that it is not simply empirical does not imply that the statement lacks content (that, of course, was the contention of the logical positivists in the early twentieth century, and they were wrong). Not all content is simple empirical content. The claim that God is creator is pregnant with the sort of content that moves humans to action, and that provides their lives with an orientation.

If it is indeed the case that God created the universe, then to acknowledge this doesn't serve as an alternative to going into the laboratory. Rather, it tells us that it's worthwhile to go into the laboratory. The statement that God is creator presents an encouragement to explore, and it holds out a promise that the universe is intelligible, and our labours to understand it are not in vain.

12

Biological evolution

12.1 What is the story of life on Earth?

Well, it is the evolutionary story, of course. The story of simple beginnings and gradual development; the story of characteristics inherited through genes, with slight adjustments that accumulate over the generations. The story of finite lifespan in an environment offering limited resources, with the consequent filtering process known as natural selection. All this can be discovered by scientific research, and it has been so discovered by all the people who joined in with the mainstream scientific community.

But what is the story of this story? What kind of a narrative do we have here? Is it tragedy? Or a comedy of errors? Or a heroic epic? Or farce? Or is it a tale of boundless exploration? Or a triumph of the aggressive? Or a triumph of the adaptable? Is it the story of brute force? Or is it the story of courage in spite of brute force? A story of increasing depth of experience? Is it a good story? Is it a story of good? Is it good?

Modern-day advocates for atheism tend to assume, and imply, or even dogmatically assert, that biological evolution is 'on their side'. Since it is a natural process, and not punctuated by miracles as far as we can tell, they infer that it stands fully apart from God, with no connection to or need of God. That inference is illogical. When it is made, it usually represents a failure to understand the relation in which God stands to the processes of the natural world. It is not easy to express that relation correctly, however. We shall attempt to say something about that in this and the next chapter. Before we do, let's at least drop the false idea that biological science is all done by confident atheists while some confused theists try to refute it, and other timid theists go along with it. There are, sadly, some confused theists who try to refute good evidence using poor arguments, but this is not the main picture. The mainstream scientific community, as a matter of fact, is made up of a mix of people. In the biology and zoology labs in most major universities you will find thoughtful Christians doing high-calibre biological research alongside their colleagues. They are a minority, but the truthfulness of their or anyone else's ideas is not measured by counting heads.

Let's suppose, for the sake of argument, that there were no miracles on planet Earth before the beginning of human history. Here, by a 'miracle' we mean a process falling sufficiently outside the ordinary patterns of nature that it could not be attributed to those patterns, nor to inspiration nor providence, nor to chance.[1] If this is the case, then of course when looking at the fossil record, and the genetic record, and the chemistry and the geology, and so on, we shall never discover, in any specific process, the fingerprint of some influence outside what could happen within the parameters allowed by the ordinary patterns of the natural world. That doesn't mean there is no meaningful sign or 'fingerprint' to discover; it means that the meaningful sign, if it is there, is expressed in step with or in tune with those patterns, and it is expressed by the process as a whole. It is not a fingerprint but a handprint.

This allusion to fingers and hands conjures up some rather unhelpful anthropomorphic images. Let's throw out those images, then, and simply ask: Does this whole process of biological history *signal* anything? Does it have any *significance*?

Various writers have attempted to express something of what they see as the possible significance here. Bertrand Russell, for example, concluded that the lesson was that impersonal, thoughtless force and chance is all you can find outside the human race, but humans somehow have more to offer.[2] Some writers express their dismay at the unrelenting suffering they see. But those same writers also often say the whole process impresses them in a rather more positive and fulfilling way, and they devote much time and energy to exploring it in detail. The suffering is there, but so is the life. The death rate is precisely one hundred percent for all organisms that ever experienced life. And yet death is not the overall winner. Life is still ongoing, after all: death has not won yet. The big picture, so far, is neither stasis nor decay but the realization of ever-changing varieties of wonderful life.

Some writers say this ever-changing variety is simply that: change, but without any sense of direction. We will argue that this is a half-truth.

---

[1] We shall see how difficult it is to give a more precise definition of miracles in Section 16.2.

[2] Bertrand Russell, A Free Man's Worship (originally published as 'The Free Man's Worship' in 1903).

The true part of this half-truth is that we don't find the idea of some sort of overarching directive force (directing towards increased complexity, for example) either necessary or useful in making sense of biological history. Nevertheless, the fossil record and the genetic record is not one of mere change. There is clear and abundant evidence of not just change but also development. Simple life forms come first, complex ones later (alongside further simple ones). You won't find any rabbits, nor any creatures with a skeleton, nervous system, and regulated body temperature, in the pre-Cambrian period. This is too obvious. But a rabbit is not just different from a bacterium; a rabbit is a richer realization of the verb *to be*. That is what we mean by *development* as opposed to mere change. The whole story is somewhat like an example of percolation: at every stage, life on Earth percolates into the available adjacent space of ecological niches, and what is available is significantly influenced by what has gone before. 'Tools' such as nuclei, cell walls, sexual reproduction, eyes, and livers are handed on and built upon or reconfigured.

At the molecular scale there is the driving force provided by the solar-powered heat engine which provides useable energy and drives chemical reactions. At the large scale there is, instead of a driving force, simply the meaning of the verb *to be*. This large-scale aspect is not a force but a space of possibility, waiting to be populated. It is the truth about the ways in which things can be, especially things that must live alongside one another. Evolution on Earth cannot go absolutely anywhere. It cannot go in all directions, completely randomly. It can only go into the space of possibility, not impossibility. It can only explore the class of all ways of living alongside one another that can persist on Earth.[3]

Conditions are sometimes harsh. In periods of climate change, large numbers of species have died out altogether, and this has happened several times. This is why people are sometimes drawn to talk of farce. Or, to take another tack, if you weigh up the biomass, you find it is mostly bacteria, then insects, with mammals and humans coming down the list a bit. So if you think mass is the main indicator of significance, then you would say life on Earth is mostly about bacteria, with some incidental complex creatures providing some of the habitat. However,

---

[3] Simon Conway Morris, *The Runes of Evolution: How the Universe became Self-Aware* (Templeton Press, 2015).

we don't think an insightful sort of person would seriously assess the overall significance or meaning of living things by counting how many kilograms they weigh. That sort of idea, if seriously insisted upon, strikes us as a sad loss of sense. In any case, this book is not addressed to anyone who thinks that (with such a person, we would have to start a long way further back).

If anyone thinks that a rabbit is no more worthy of attention than a dead piece of fur flapping in the breeze, then this book is not addressed to such a person.

Sometimes people feel that questions about meaning are not well-posed questions. Perhaps they are non-questions. If you found some random scratches on a stone, it would be simply a waste of time to ask 'What do they mean?' in the sense of some sort of written message. That would be an example of a category error, or a failure to understand what sort of thing you are dealing with. The only 'meaning' such scratches have is that something randomly scratched the stone. But if you found the Rosetta stone, then you would be right to ask what the scratches on it mean.

Asking about the meaning, if there is one, of life on Earth is very close to asking about the meaning, if there is one, of the whole universe. It is right to be hesitant about volunteering answers to questions as big as that. But we can't help but answer, up to a point, because we must choose how to live. What we do today is our answer, whether we like it or not. We detect what we think matters in the world around us, and we contribute to that. This is what each of us does.

This is no small issue, either. We talk of things going *wrong* when someone suffers some affliction, a cancer or a famine or an enslavement. We say we are trying to make things *right* or *better* when we study medicine or develop agriculture or liberate slaves. These are all judgements about value and significance. They are saying that life on Earth is not a farce. They amount to saying that we see ourselves as some sort of trustees, who think there is something worth preserving and improving, and we have volunteered for the job of preservation and improvement. Consider, for example, the situation of starvation for some while others have plenty, as compared to the situation of equal shares for all. These two situations are not just different ways that things happen to fall out. One is objectionable, one is not.

The only appropriate answer that a human being can give to the question 'Is the story of life on Earth a good thing?' is to respond,

'How can I possibly pass judgement on such a question? My resources of knowledge and wisdom are utterly inadequate to take everything into account or even to know what should be taken into account. I can only acknowledge that I am part of this process; that, for good or ill, it gave birth to me, and my role is to play my part, and make it a part for good: for making the process better.'

Gerard Manley Hopkins put it this way in his poem 'As Kingfishers Catch Fire, Dragonflies Draw Flame':

> Each mortal thing does one thing and the same:
> Deals out that being indoors each one dwells;
> Selves—goes itself; *myself* it speaks and spells,
> Crying *Whát I dó is me: for that I came.*
>
> I say móre: the just man justices;
> Keeps grace: thát keeps all his goings graces;

Hopkins is saying that the man who is just puts his attitude into practice; he becomes a means whereby justice is established. Instead of passively voicing abstract answers to abstract questions that are too large for us, we must play our part as best we can. Most of us feel that, even after looking squarely at all the parasitism and pain, the mass extinctions and the locust swarms and the predation, nevertheless this story of life on Earth is tremendously valuable, and even, yes, good. Every willing parent has voted that it should continue.

Now let us return to the question: What has God got to do with it? To expand a little on the operational meaning which we gave in the glossary at the end of Chapter 3, by 'God' we mean the reality both that sources the universe with its own integrity as a world and that also continuously relates to that world, on every level. Not by overturning it, but by working with it. An atheist will want to say that that reality is something strictly abstract and impersonal, such as 'mathematics' or 'the laws of nature'. The theist asserts that there is more to God than that. We are not talking about anthropomorphic super-beings. We are talking about the human experience of *being known*, and the experience of *sharing the world*. We show the world to God (how it looks to us), and God shows the world to us. Sensitive and thoughtful people have told us that our truest parent is one that understands the nature of poverty and suffering as well as the nature of joy, and these people seem to be among the wisest and best among us.

Our genetic inheritance provides us with complex brains that can operate rationally, but by itself it is not enough to enable us to think and act well. Nothing in evolutionary biology leads us to believe that genes make us good. At the level of inherited nature, humans are not, deep down, innately committed to good, even by our own standards. Deep down we are thoroughly compromised, and must make an effort at good, finding the strength to do this, if we ever do find it, somewhere other than in our genes. (This truth about human nature is plain enough from observing what humans do, and it was asserted by Christian thinkers well before the modern era; evolutionary biology has added some confirming detail.) But that other component—the one that does not come from our genes—is also at work in human history, and something like it has been at work in biological history. In unconscious living things it is an astonishing, continually reasserted, deep resource of beauty somehow woven into the very fabric of the world. In conscious living things it is a hunger for truth and goodness, a hunger that is met, deep in who we are, through a process we do not comprehend, but whatever is the reality at the source of it, that is our truest parent.

What we have written in this section is an opening gesture, a rapid survey of how one may get to grips with the issues raised by the story of life on Earth. To say that the human race has been created by God and to say that the human race has come about through a long sequence of natural processes is not to say two mutually inconsistent or contradictory things. Nor is it to add a superfluous colouring to a picture which does not need it. Rather, the religious statement indicates the role and significance of the natural process. In the rest of this chapter and the next we will unpack these issues at greater length.

12.2 Consensus on the scientific details

There is circulating in popular culture, and in some parts of academic culture, the ungenerous idea that the neo-Darwinian synthesis somehow belongs to atheism.

No, it does not.

The modern evolutionary synthesis or scientific account of the development of life on Earth is part of the communal intellectual property of all humankind. Don't you dare declare exclusive ownership of it. It no more belongs exclusively to atheism, nor to theism, than does the air we breathe or the fact that two plus two equals four.

The first thing to say about the development of life on Earth is that we can agree on the scientific details, and the way to agree on them is to join in with all the honest efforts to learn and to correct mistakes that go on in standard scientific work. That is, we recognize the work of the ordinary, mainstream scientific community as the most careful, intelligent, hopeful, and reliable source of knowledge about the details of physical processes. By this we don't mean to imply that that community does not make mistakes or need to keep discussing issues with people in other walks of life. We mean simply that it is within the ongoing discussions and presentations of ideas and evidence in the worldwide scientific community that we have the best hope of moving forward together in our scientific understanding.

We reject the recent set of ideas called 'Intelligent Design' or ID for reasons we shall articulate in the final section of this chapter. We will briefly summarize our reactions to this set of ideas here, but we do not want this to distract from the main flow of the present chapter, so the more lengthy discussion is in the final section, which the reader can omit if preferred. Our conclusions are as follows. First, ID proponents are correct when they point out that most evolutionary pathways have not been uncovered in detail, but the significance of this should be assessed in a scientific manner. In view of the richness and complexity of life on Earth, how many of the evolutionary pathways should we expect to have discovered in the time available since this area of science became well organized? The mainstream scientific community is, we think, offering a reasonable assessment of this. Next, a major component of ID arguments, called 'irreducible complexity', does not imply what the ID argument says it implies. Finally, ID has become rapidly associated with a social and media movement which has supported dubious practice in public presentation of science and history (Section 12.5 clarifies this accusation, which we do not make lightly).

From the point of view of large parts of the Christian community, the ID movement appears to be a distraction. In so far as it does not adopt the sort of caution and careful argument that is characteristic of science, it cannot be helpful to other concerns such as theology. The attitude of mature theology to evolutionary science was well summarized by philosopher and theologian Keith Ward:

> Most theologians accept that whatever view we take of the physical universe has to be consistent with evolutionary science.[4]

---

[4] K. Ward, *A Vision to Pursue: Beyond the Crisis in Christianity* (SCM Press, 1991).

The consensus in science is always a developing consensus. It is always in need of refinement and sometimes in need of large correction. So we acknowledge the right of any investigator to propose a largely different way of understanding any given area of scientific investigation. We keep seeking. However, all such proposals must submit to the court of rational consideration in the established way. We require evidence to be accurately reported, phenomena either to be repeatable in the present or to exhibit a convincing ensemble of past examples (the quality required of that ensemble depending on the weight of the claims being based on it). We require that the idea should lead us to expect something, or enable us to look for some effect, and then that the effect is indeed observed. We require that the idea does not unnecessarily multiply hypotheses, and we look for some sense of insight. The last two components are as important as the rest. They are part of what science cultivates and prospers by: the aesthetic quality that ideas can have: an innate simplicity or 'just-so-ness' that makes them useful components of an intellectual structure.

At the time of writing, the standard account of evolution by inheritance with modification and natural selection—the commonly adopted neo-Darwinian synthesis—has all these properties, and we know of no aspect of the development of life on Earth that could not be accounted for this way, except possibly the origin of life on Earth. While the essential ideas remain robust, evolutionary biology has never been a more active field of research. Many of the current advances have to do with epigenetics (literally 'on top of the gene'). This is the study of how chemical modifications of DNA and its surrounding proteins regulate how the genes are switched on and off. Epigenetics explains how two cells with the same DNA know that one should be part of a nose and the other should be part of a liver. New discoveries are showing that, to some limited extent, acquired characteristics may be inherited.[5] Such inheritance is reversible, but it can propagate across generations. It is even possible that parental circumstances and choices, such as stress and diet, may have effects on the early growth of a child which are passed on epigenetically. We may thus discover closer connections between physiology and evolution than were previously supposed. Although we can hope that such advances will give new insights into

[5] D. Noble, 'Physiology is Rocking the Foundations of Evolutionary Biology', *Exp. Physiol.* **98** (2013), 1235–43.

the mechanisms of evolution, we can remain reasonably confident about the current picture of the historical sequence of evolution.

The question of the origin of life remains open; we do not know when life first followed a broadly speaking 'Darwinian evolutionary' process at the interface of chemistry and biology. The origin of life, at least in its early processes, could not strictly speaking be Darwinian before there were units of replication that could undergo variation. In any case, we think it perfectly possible that the first forms of life arose by natural processes following their normal characteristic patterns.[6] Such processes are in principle reproducible and not miraculous in the ordinary sense of the word.

Christian reflection need not, and in our opinion should not, see the natural development of the world as a difficulty or a counter-evidence which must be doubted or somehow compensated for. On the contrary, it is something to be expected and celebrated in a sympathetic and wise manner. When we share our experience of God with each other and the wider world, we are sharing the conclusion that through this very development a transcendent purpose is being realized, in which we are invited to join. We look to God not because we failed to see how something could have happened, but because we begin to see *what is was that happened*, in the sense of what good is being realized through it.

The aim of the next chapter is largely to take up that challenge of seeing, insightfully and sympathetically, what it was that happened in the case of the story of life on Earth. Before doing so, we need to clear some misconceptions out of the way. We don't want to eschew Darwinian evolution, but we do want to expose misconceived glosses on it.

12.3 Hagiography of Charles Darwin

Darwin's work, *On the Origin of Species*, is a remarkable intellectual achievement (and, we might add, a remarkable literary achievement), but the theory of evolution should not be presented as if it is all due to Darwin. The ideas he discovered were ripe for discovery; significant elements, such as the key idea of natural selection, were independently discovered by Alfred Russel Wallace, and helpful contributions were

---

[6] A. Wagner, *The Arrival of the Fittest* (Current, 2014).

also being put forward by others in the scientific community.[7] Even Darwin's own grandfather Erasmus Darwin had made useful contributions. Darwin published a fair assessment of this prior work in his preface to the third edition of *On the Origin of Species*. And, of course, Darwin did not get everything right. His views on the mechanism of inheritance were largely wrong and the work of getting the right view of this is due to Gregor Mendel and others. To say this is not to denigrate Darwin's contribution. It is to be fair and correct, as indeed he was himself.

Like the majority of commentators, we are tremendously impressed by Darwin's life achievement in science; he holds a high place in the history of science and merits his recognition on earlier British ten-pound banknotes. Nevertheless, his work takes its place in the normal way as part of a communal effort from which he benefitted and to which he contributed.

The reason one should not present evolution as if Darwin arrived at it single-handedly, or as if no one else could conceive of it at that time, is partly because both claims are untrue, and partly because it undermines general scientific education in the present. It suggests to members of the general public that this area of science is subjective, a matter of one person's opinion rather than a consensus arrived at by the accumulation of evidence and understanding, and since built upon and more firmly established by large numbers of other contributions. We would argue that evolutionary biology should be seen as both a well-established set of ideas, and an ongoing research discipline.

A related mistake is to overstate the contribution which the understanding of evolution makes to wider philosophical questions. It certainly makes a very large contribution, but the situation is not quite as simple as various writers have suggested. For example, the zoologist G. G. Simpson, when considering the question 'What is man?', stated: 'The point I want to make now is that all attempts to answer that question before 1859 are worthless and that we will be better off if we ignore them completely.' This sweeping statement seems daringly grand and bold at first, but one should certainly think a bit before buying into it. The answers provided by poets from at least as early as Homer (the writer associated with the ancient Greek epics the *Iliad* and the *Odyssey*)

---

[7] Philip Ball, 'If Not Darwin, Who?', *Nautilus* (15 December 2016); available at http://nautil.us/issue/43/heroes/if-not-darwin-who.

and by the writers of the Psalms, and by ordinary folk down the ages as they sought peace and tried to help the next generation, have included some pretty good working answers which certainly should not be ignored.

12.4 Explaining the mystery of existence?

In his books *The Selfish Gene* and *The Blind Watchmaker*, Richard Dawkins presents a 'gene's eye view' of biological evolution. Both books are eloquent, lyrical, immensely persuasive, and insightful about the processes of Darwinian evolution. However, they also adopt some misleading and unsustainable positions which we will comment on here.

These books suggest to the reader that faithful followers of Jesus of Nazareth still hold a view expounded by a scientifically literate clergyman in the eighteenth century, William Paley (whom we mentioned briefly in chapter 3). Both books correctly show how a central part of Paley's idea is superseded by the mechanism of evolution. The misleading suggestion is that this means it also overturns the rationality or sound good sense of Christian commitments in general. No such thing is true.

William Paley was working more than half a century prior to Darwin's ideas being expounded and generally accepted. Paley's book *Principles of Moral and Political Philosophy* was hugely influential. It was cited in debates in the UK Parliament and in America's Congress. It was required reading for Darwin as an undergraduate at Cambridge, but Darwin said that he learned most from another book by Paley, with the long but informative title *Natural Theology: or, Evidences of the Existence and Attributes of the Deity; Collected from the Appearances of Nature*.[8] This is the book in which Paley compares the structure of ecology with the structure of a watch, and argues by analogy that they must have come about by similar processes. This is now recognized to be a faulty argument. However, the book presented the argument through a wealth of well-observed examples, and these are what both informed and fascinated Darwin.

Darwin followed Paley as a student at Christ's College, Cambridge, and their portraits face one another in the dining hall.

Paley's work, like many other intellectual contributions of his time, promoted progress in science even though one of his main points was wrong. Richard Dawkins would not want to deny this, but in order to

---

[8] W. Paley, *Natural Theology: or, Evidences of the Existence and Attributes of the Deity; Collected from the Appearances of Nature* (Philadelphia, 1802).

engage with the philosophical and theological implications of zoology one must address the way they are thought about now, after Darwinian evolution has become widely understood, not the way they were thought about by William Paley. This is what Dawkins fails to do, except in a very limited way. He brings in ways of thinking that modern-day atheism might wish to affirm, and simply ignores other well-informed and thoughtful contemporary voices.

In essence, Dawkins presents the evolutionary process as an argument for atheism. But the analysis and description of a process cannot, logically, even address the issue of the overall meaning and purpose of that process, nor can it address what made it possible for that process to happen in the first place. It follows that no such analysis addresses the relative merits of atheism and theism, properly understood. The impression that this analysis does so is therefore false.

The unsustainable position is the exaggerated language that Dawkins adopts when he presents what he thinks he has achieved. For example, in the preface to *The Blind Watchmaker* he writes that the book has been written 'to persuade the reader, not just that the Darwinian world-view happens to be true, but that it is the only known theory that could, in principle, solve the mystery of our existence.' Solve the mystery of our existence, no less! This misdirects the reader, because of course Dawkins has the right to say whatever he likes about what he hopes to persuade the reader of, but it does not follow that this is what will be achieved. He then presents a persuasive case that living things did indeed develop by the mechanisms called Darwinian (yes, well we knew that). We are left with the impression that the expressed hope has been realized. At the end of the book, there is no addendum saying,

> so Darwinian evolution is what happened; there's no good reason to doubt it. But I'm sorry I didn't manage to show that such arguments could in principle solve the mystery of existence for you, as I had hoped. I wonder why the natural world has these long-lasting physical patterns that make chemistry possible and that underpin biology. I wonder why mathematics gives a good working language for physics.[9] I wonder why the universe has the combination of durability and richness that makes life possible. I wonder what it is about Keats' poetry that makes him a greater poet than Ezra Pound. I wonder why people find mere propagation

[9] E. Wigner, 'The Unreasonable Effectiveness of Mathematics in the Natural Sciences', *Commun. Pure Appl. Math.* **13** (1960), 1–14.

of genes to be a pitifully inadequate account of personal identity and of what we aspire to realize in our lives.

Perhaps we are being ungenerous here and should allow an author a bit of hyperbolic writing in an enthusiastic introduction. But it is not presented merely as an overstated introduction. It is part of the main argument. He writes, for example:

> Natural selection, the blind, unconscious, automatic process which Darwin discovered, and which we now know is the explanation for the existence and apparently purposeful form of all life, has no purpose in mind.

Note that phrase 'we now know'. It is a misdirection. We do now know that Darwinian evolution is what happened. But it does not follow that we now know what Dawkins says we now know. The Darwinian process 'is the explanation for the existence...of all life', he writes. This is unsustainable. The most one can say is that this process is *part* of the explanation, because no physical process can explain its own existence, and furthermore, this is the really central point here, *no process of exploration can explain what is found by the exploration.*

The actual logic of the situation has these two further ingredients, and they are not incidental; they are large and central. The argument of *The Blind Watchmaker* fails to mention either. On the first: the argument can only proceed if one takes as a 'given' that a stable and rich physical universe is in continuing existence. That is, it is required that there is a large amount of material following patterns captured by the relevant mathematics and with the right degree of stability, complexity, and slow change, etc, etc. The logic has the following form: if this is so, then a Darwinian type of development can take place. Therefore, the theme of modification, descent, and selection is not 'the explanation' but only a small part of the explanation for life.

This first ingredient, that an essential mechanism is not the same as a complete explanation, might not trouble someone sympathetic to Dawkins' case, perhaps; one could simply add the remark that whatever the further causes are, according to Dawkins they may safely be assumed to be thoroughly impersonal. The second further ingredient is, however, equally important and more telling. It is that the structures and behaviours of the organisms that come about are not arbitrary. Darwinian evolution cannot lead to just any type of ecosystem; it can only lead to ecosystems in which there is sufficient negotiation among organisms, so that they do not obliterate one another in mutual assured destruction. That negotiation itself is subject to further rules and patterns that the evolutionary process does not invent but can only respect. It is not owing to natural selection that round bones are more efficient than square ones—this is owing to materials science and engineering. Natural selection favoured animals that embodied this truth. Similarly, it is not owing to natural selection that mutual trust is better than mutual suspicion—this is owing to the nature of personal relations. We are not here saying that organisms inclined to trust one another could not come about by a Darwinian route; we think that they can, just as round bones and trunks can come about by a Darwinian route. But that is not the point here. The point is that the role of trust and suspicion in personal relations is what it is. No process, Darwinian or other, influences this any more than it can influence whether round or square shapes are the more efficient. We are speaking about the very concept of personhood. Natural selection, as it were, discovered this; it does not cause it. Natural selection favoured the eventual emergence of complex creatures able to embody personhood; when this emerged, it could not do other than embody what personhood is. The 'mystery of our existence' is, in fact, very much about the nature of personhood. The nature of personhood is not explained by the physical process through which it became embodied in the physical world.

To see this, consider a simpler example. As an end result of the evolutionary process, we possess brains that can perform elementary arithmetic such as '2 + 3 = 5'. But it does not follow from this that arithmetic itself is the product of genetics and natural selection. That is like saying, 'You only believe that because you are a mathematician.' If some other species developed to the point of being able to understand arithmetic, then they will agree that two added to three makes five. If some other species developed an understanding of something more complex, such as the geometry of flat spaces (called by us Euclidean geometry), they would not think that the angles of a flat triangle add up to something other than 180°. This is because Euclidean geometry has its own rules and structures that are innate to it and are nothing to do with evolution. In a similar way, our brains give us access to things such as justice and friendship, but biological genetics does not and cannot influence, even by one iota, what such things really are. Our biology can only furnish our ability to be aware of and responsive to these realities.[10]

[10] This point is carefully argued and presented at length in Andrew Steane, *Science and Humanity* (Oxford University Press, 2018).

To repeat: the innate nature of friendship is not caused by anything in biology. All that biology can do is influence the ability of biological forms to embody friendship. And similar things can be said about much else that is significant to human identity. We shall illustrate further how an effect can transcend its immediate causes with an analogy of an arched bridge in Section 13.1.

Now let us return to the quotation from Dawkins and peruse one more item.

Note the phrase 'apparently purposeful' in the quotation from Richard Dawkins. It seems to suggest that purpose is an appearance only. This too is misleading. There is no good reason to doubt that human lives are purposeful. And there is no good reason to doubt that the whole process of biology is itself there for a good purpose. The purpose was to achieve the very thing that has been achieved: the huge variety of inter-dependent and increasingly deep and rich life, largely free to discover what life is. But our basis for saying that is theological rather than scientific.

Professor Dawkins' eloquence has been motivated by a passionate enthusiasm for science, and he has greatly desired to advance the public understanding of science. But this has been coloured by his passionate enthusiasm for the propagation of his atheism. In consequence, and this is a sad irony, Professor Dawkins' efforts have probably done as much harm as good in the public understanding of science. His work has served to harden in the minds of his readers the notion that one must choose between scientific integrity and theism, as if they were mutually exclusive alternatives. This is a false choice. The truth of the situation is not mutual exclusion but mutual enrichment. There are versions of theism which cannot accommodate evolutionary science in full, and there are versions which can and do. The latter are main-stream versions, which ever since Darwin's time have received carefully articulated support from intelligent people.

In view of the fact that these are important matters which touch on education policy, intellectual integrity, values, and aspirations about how life is to be lived, they should be presented in an open way that does not deny or ignore the breadth of religious opinion.

12.5 On the 'Intelligent Design' movement

We finish this chapter with a discussion of the arguments which have become known under the title of 'Intelligent Design'. This section can be omitted by readers not interested in that topic, though it provides something of a historical case study of how such issues can be addressed.

The phrase 'intelligent design' can refer to a number of different statements or ideas; in this section we mean by it the set of ideas championed by scientists such as Michael Behe and the mathematician and philosopher William Dembski. Michael Behe introduced the group of ideas known as Intelligent Design in his book *Darwin's Black Box* (1996), with much emphasis on a concept he called 'irreducible complexity'. In brief, ID is the claim that there are various biological structures or processes that could not have come about by the gradual process that Darwinian evolution involves, and which therefore must have come about by another process. The process suggested by ID is one which may be called 'intelligent' because it involves marshalling together a number of disparate parts at a rate that is far above what could happen by chance, with the end product functioning through the orchestrated effect of all the parts, such that there is a reasonable inference to design. That is, the end product is like other end products that we know to have come about by intelligent agency, namely human artefacts, and the inference is sufficiently strongly suggested as to make it reasonable to adopt.

This argument immediately strikes a typical modern scientist as much like the arguments that were made before the nineteenth century about the evidence for design in specific developments of the natural world. Since those previous arguments were largely unsound, one is naturally wary that this is a repeat of a previous mistake. In the USA, this whole area is bound up with political issues surrounding education policy. This makes the strictly scientific discussion veer rather quickly into defensive modes of expression, and not all of it is careful and fair (on both sides). Nevertheless, ID has been given a fair hearing. It was taken sufficiently seriously that an effort was made to examine the examples used by Behe to make his case.

It quickly emerged that there is a very great problem with the ID concept called irreducible complexity. The word 'irreducible' here refers to the idea that a structure or process might consist of a number of parts, such that the structure or process can only work if all the parts are present. Behe's example is that of the flagellar motor present in some bacteria, or the immune system in mammals. For such a structure or process, the removal or absence of any one part would suffice to render it inoperative. Behe claimed that such machines or processes could not arise by gradual accumulation of small changes because the incomplete structure, before the final stage was in place, could not be functional. However, this is certainly wrong. It is possible to build a machine gradually such that it has some working function at every stage, and yet the final machine requires all its parts in order to work. Sometimes the last stage can be to remove a component which the machine previously relied on, such as a support, but which it does not need once it is complete, as long as all the other parts are now present. The notion of irreducible complexity does not capture a concept that is relevant to understanding biological evolution. This explains why it has not caught on in the mainstream scientific community.

Let us, nonetheless, try to set out an ID approach that might make sense. We are not at this point asserting that any individual person such as Michael Behe advocates any particular idea; we are simply going to present what a case for ID might look like if it were to merit further study.

Such an ID approach will agree with many of the facts on which Darwin and subsequent workers based their ideas. The following are so well established as to be beyond dispute (because the evidence is overwhelming), and the present dispute need not question them:

- Living things tend to produce many more offspring than survive (think of the seeds of the sycamore tree, or frogspawn, for example).
- Traits are mostly inherited by passing on information-bearing molecules (the DNA), though some epigenetic information can also be passed on.
- Each new offspring inherits DNA from its parents, and in addition to new combinations this can include small changes called mutations which happen randomly in the sense that they occur without the well-being or otherwise of the organism in view.
- Offspring whose inherited characteristics tend to help them live to reproductive age and produce more offspring will for that reason pass on their genes to more progeny, a process known as 'reproductive success'.
- The time scale is long; so long, in fact, that there have been enough generations for these facts to be themselves enough to account for

much of the gradual and beautiful story that has taken place in the development of life on planet Earth.

Now one problem is that some commentators have wanted to take this wonderful story and add to it another ingredient: atheism. Because there are no miracles here, they say, there is no evidence for God. But we say that the whole process is one big miracle.[11]

However, science remains open and the human race has not arrived at perfect knowledge of anything. This is why we left a carefully chosen 'chink' in the last item of our statement of what is 'beyond dispute'. We said that it is not disputed that the neo-Darwinian model is enough to account for 'much' of the story of life on Earth. There remains the open question: Is it enough to account for not just much but essentially *all*? Have *all* developments followed the route of gradually accumulated small change, with each such change open and undirected? This is a question to which no one can ever know the answer for sure, but one can modify the question to:

> In the context of evidence accumulated to date, is it reasonable to believe that all developments, after the process began, have followed the 'Darwinian' route?

We (the authors) think the answer to this question is 'yes'. We do not know of an example for which it has been shown that the neo-Darwinian model is sufficiently dubious as to be overthrown. Whenever someone has suggested such a counter-example, further work has shown that, in fact, the supposed 'counter-example' does show the features one would expect within the Darwinian paradigm. Examples of such features are the presence of opportunistic making-use of existing structures; intermediate stages that each themselves function in some fashion or are neutral; enough time for these stages to accumulate.

However, it remains true that the natural world is full of marvels and it is very easy to find examples of existing structures or systems for which the details of an evolutionary pathway are not known and are hard to guess. So Behe is not only within his rights to say this, but also right to say it (from which starting point he makes unwarranted

---

[11] We have in mind here both parts of the theological term 'miracle'. That is, we consider that the inauguration and ongoing support for the physical world as a world with its own integrity is owing to God, not an intermediate cause, and we consider that the result has remarkable significance. We shall evaluate such ideas further in Chapters 16 and 17.

inferences, but the starting point itself is warranted). The problem lies in what may be deduced from this observation. The fact that one cannot see how a given system could have travelled gradually to a given end point does not imply that the journey was not gradual. It might simply be hard for us to work out, a case of human ignorance. To decide this, one looks at what is known and one tries to come to a reasonable overall assessment.

The mainstream scientific community has done this. That is, people have been reasonable. ID has not been dismissed out of hand (though we already admitted many are wary of it and would require a strong case to be persuaded by it). Some people do write very dismissively and with bad grace, but some have patiently examined the case. The result is that ID is not dismissed out of hand, but it is dismissed. That is, it is found, after scrutiny, to be lacking in coherence for some of its basic ideas and lacking in evidence for its claims.[12]

We already mentioned that one of the basic ideas, namely irreducible complexity, is incoherent, in the sense that it makes either an untrue or an irrelevant statement. This is enough to show that no provably or self-evidently non-Darwinian structures have ever been discovered. The ID concept might still be preserved if a case could be found where the Darwinian process was highly dubious.

The most famous of the examples put forward by Behe has been mentioned already—the bacterial flagellar motor. It was claimed that it could not come about by numerous, successive, slight modifications of what went before, because an intermediate form would not be functional and therefore would not be selected for. Now this is quite a complex molecular machine and details of how it could come about gradually were discovered only after Behe had written his book. A functional smaller machine has been identified, one which has many of the parts of the complete motor missing and does a different job. Also, most of the proteins involved in the molecular motor have similar forms that are used in cells for other purposes. All this is what one would expect in the Darwinian model. In response, ID proponents claimed that this intermediate structure would itself be irreducibly complex, or that it came after not before the full motor. We suspect that they are wrong on either the first or both points, but in any case

---

[12] D. R. Alexander, *Creation or Evolution: Do we have to Choose?*, 2nd edn (Monarch, 2014), Chapter 14.

it is clear that the complete flagellar motor does have functional intermediate stages.

Another telling story emerges from the consideration of the immune system in mammals. It was suggested that no Darwinian process could result in this complex system, but when one looks into the details of the structures involved, they do have the sorts of properties that one would expect on a Darwinian model. There are various ways in which simpler versions of the system can still be functional, and one can propose sequences of stages whereby the whole system could have come about gradually.

In view of the richness of the biosphere and the huge number of complex biochemical processes, and the limited amount of human time and insight, one would not expect any but a small fraction of the evolutionary pathways to have been worked out in detail by the scientific community, and this is the case. Seeing this, Behe, in *Darwin's Black Box,* proceeded to the claim that, in fact, none have been so traced! This claim is an overstatement. It is a case of what level of detail one may reasonably expect to have achieved in the time available for research. A good case study is provided by the Krebs cycle. This is a biochemical pathway consisting of nine enzymes and a number of cofactors; it plays a central role in cellular metabolism. It is a real, complex, biochemical system whose Darwinian development was unknown and hard to envisage when the details of the system first became clear. However, evidence for a Darwinian account steadily grew, to the point that in 1996 a study could conclude the following:[13]

> The emergence of the Krebs cycle has been a typical case of opportunism in molecular evolution.

Subsequent work that brings in DNA sequencing has confirmed the validity of the model outlined by those authors. A proposal for using this example in an educational context has been put forward.[14]

---

[13] E. Meléndez-Hevia, T. G. Waddell, and M. Cascante, 'The Puzzle of the Krebs Citric Acid Cycle: Assembling the Pieces of Chemically Feasible Reactions, and Opportunism in the Design of Metabolic Pathways during Evolution', *J. Mol. Evol.* **43**:3 (1996), 293–303.

[14] Caetano da Costa and Eduardo Galembeck, 'The Evolution of the Krebs Cycle: A Promising Subject for Meaningful Learning of Biochemistry', *Biochem. Mol. Biol. Educ.* **44**:3 (2016); available at http://onlinelibrary.wiley.com/doi/10.1002/bmb.20946/pdf.

This example illustrates the fruitfulness of the working assumption that there have been no abrupt and localized infusions of information into the development of life on Earth. If one assumes such abrupt infusions have occurred, then at any point one might simply give up looking for more understanding, and just say 'here is the influence of the intelligent information-giver or designer.' If meanwhile your colleagues decided to keep looking into the structure in question, and they discovered more and more about how the structure functioned and how it may have come about, then they are the ones who are bringing the learning which we all want, and whose efforts merit our support.

The case for Darwinian evolution holds good if there are enough generations for the gradual route to the end product to be followed, and if there are steps along the route that can work. The ID movement points out that there are plenty of examples where we really do not know for sure that there exists such a route. However, there are enough examples where the Darwinian process is known to be sufficient, and enough hints that it also works for cases still being worked out, that there is not good enough reason to adopt ideas such as ID. That, at least, is a reasonable opinion and it is our opinion. Behe is not ignorant of evolutionary biology, and the reason his book has been influential is because it includes a reasonably large collection of detailed examples, but nevertheless his claim that 'no' evolutionary pathways have been worked out for real complex biochemical systems is sufficiently overstated as to be misleading.

Here we have focused on some biological examples put forward by Michael Behe. We will not devote space to further aspects of ID, such as philosophical points advocated by Philip Johnson and William Dembski, except to say that we are unconvinced by that aspect too. Their arguments have much in common with the type of approach which we critiqued in Chapter 11.

ID rapidly became a political and media movement. The movie *Expelled: No Intelligence Allowed* (directed by Nathan Frankowski) presents a highly misleading picture of the history of the movement, and a distorted picture of both the modern scientific community and of science itself. Overall, one may suspect that in the Intelligent Design movement there is something other than straightforward desire to discover what is so; it is more of a social phenomenon than a contribution to improved understanding.

Long before the ID movement was even conceived, Charles A. Coulson, a distinguished scientist and chair of the charity Oxfam, wrote, 'When we come to the scientifically unknown, our correct policy is not to rejoice because we have found God: it is to become better scientists.'[15] We concur.

[15] C. A. Coulson, *Science and the Idea of God* (Cambridge University Press, 1958), 16. Coulson (1910–74) was Rouse Ball Professor of Mathematics at Oxford University before becoming Oxford's first Professor of Theoretical Chemistry. A Methodist lay preacher, he discussed the integration of scientific and religious views in a series of BBC broadcasts, as well as in his written work such as *Science and Christian Belief* (University of North Carolina Press, 1955).

13

This is the story of life on Earth

In the previous chapter we aimed to do two things. First, we wanted to make clear that we consider it both sound and creative, and also our intellectual duty, to receive and accept the well-argued case for mainstream science in the area of evolutionary biology. Secondly, we aimed to show that those very ideas are not well interpreted by some of the widely read commentators in this area, and, more generally, we indicated some of the ways in which mistakes can be made.

In the present chapter we will move to a more positive approach. We will present an attempt to respond to what is found in evolutionary biology as honestly and correctly as we can. We won't claim that we do this from a totally neutral perspective, as if we had no prior commitments whatsoever. To claim that would simply to be unaware of the truth of the fact that no human ever approaches any large question entirely neutral and unbiased. Our bias is towards the notion that life is, in the end, not valueless nor absurd but valuable and meaningful.

What we must do in the case of evolutionary biology is not be embarrassed by it, but make ourselves the ones who see it most clearly and correctly. We can take a long, careful look, and we do not need to buy into perspectives that are not well argued. Sometimes it requires intellectual effort and philosophical maturity to distinguish what is well justified and what is not, among the large-scale claims that are made about evolution. For example, the biological fact that is indicated by the phrase 'survival of the fittest' does not necessarily imply that strength and aggression are always promoted; evolutionary 'fitness' can also be about things like adaptability and negotiation. An animal such as a tree frog might be well fitted to its ecological niche if it negotiates, in a rudimentary manner, with other tree frogs, rather than attacking them, when territorial issues arise. Also, in the mathematical facts that underlie genetic processes, the metaphor of selfishness may or may not be sound. Similarly, the metaphor of blindness is apt or not, depending on what is meant by it. And so on. There is much pain in the natural world, much disease and starvation and killing. But there is also much harmony, depth, and balanced, rich ecology.

13.1 An effect can transcend its immediate causes

Suppose that a certain biological structure has come about by a process of gradual accumulation of small changes, with each such change open and undirected. Does it follow from this that the resulting structure has no further meaning or value, beyond the fact that it is the product of such a process? Of course not. Darwinian accumulation of small change over many generations can result in both an ant and an antelope, but this does not imply that an antelope is the same as an ant, and furthermore it does not follow that an antelope is the same sort of thing as an ant. They are similar in some respects (they share a part of evolutionary history; they move about; they search for food and eat it; they live in groups; etc.). But an antelope is a member of a species with quite advanced cognition and a rich social life, whereas an ant is much simpler. So much so that we may, quite rightly, destroy a colony of ants that invaded a kitchen and brought there the risk of disease, whereas if a colony of antelope invaded our house, we would shoo them away, and hesitate to kill any, partly because we owe a greater duty of care to these creatures in and of themselves. We will not discuss the moral philosophy involved here; suffice it to say that these everyday conclusions point to the fact that the living organisms on planet Earth have much value in and of themselves, quite irrespective of the details of the process by which they came to exist.

In order to illustrate what is going on in this example, let us consider something simpler.

Consider what happens when an ordinary arched bridge is made from brick or stone. A difficulty faced by the builders is that the arch, which will eventually act as the support for the bridge, only functions once it is fully in place. During the construction of such a bridge, while the arch is being built, the bricks or stones must be supported another way. A common method is to build a framework or template from wood and lay the stones on top of this frame. Such a frame is a temporary measure, and it may be considerably less strong than the eventual stone arch that is the end product.

When the keystone is finally placed and the arch is complete, all the stones press against one another and the whole structure can support itself without the need for the wooden frame. The frame is removed, and the bridge stands forth on its own.

This example illustrates what we mean by the title of this subsection: 'an effect can transcend its immediate causes'. In our example, the immediate causes which give rise to the eventual bridge are such things as the stones lying on the frame, the frame itself, and the gradual accumulation of more and more stones. These stones never display the overall structural integrity that is a prerequisite, almost a defining property, of a bridge, until the keystone is in place. However, once the arch is complete, then it gains that structural integrity. This is a property that is true of the end product or effect (i.e. the completed bridge) but which is not true of the immediate causes, i.e. the state of affairs amongst the stones earlier in the temporal sequence. There is some structural integrity present during the construction in this example: it is to be found in the wooden frame. But the process is one in which the stones themselves only gain this property at the end, and then it is a property of the ensemble.

This example of a bridge is an apt example that captures one aspect of what has occurred time and time again in evolutionary history: the coming together of parts to make a whole that can function in new ways. The point we wish to make is not about whether such coming together falls outside the standard paradigm of random variation and natural selection. Let us assume for the sake of argument that such processes all fall within the standard 'neo-Darwinian synthesis'. (For example, in a biological case, the intermediate steps might have beneficial effects in other ways, and thus they can accumulate.) Our point is simply that a bridge is a bridge, no matter how it was made. The 'mystery of its existence' is about what qualities are true of a bridge; what aspects of reality are revealed in it; what does it embody. A bridge embodies the property of strength under compression in the vertical direction. This property is not owing to the process of gradual accumulation of stones, because the bridge would still have the property if the stones had been brought to their locations some other way. This property of compressive strength is owing to various aspects of physics, materials science, mathematics, and engineering. Similarly, the bridge has the property of making a connection from one place to another. This again is a statement about the end result which does not depend on how the bridge was brought about. All these 'mysteries' are not contained in the construction process. And the way they operate, and the reasons they work, have not much to do with the construction process. Rather, both the end product and the construction process exist as part of a grand display in which truth of richer and richer kinds can come to be embodied and enacted.

We have now presented much of the argument that the products of the evolutionary process can be rich in meaning, and the process does not undermine this conclusion in the least. By mulling on this we hope that the reader may also begin to see why the natural process called Darwinian is not threatening to theism and the building of a life with God. The life of looking to God is not one which desperately looks for odd events which can't be placed in a scientific picture. To seek to know and serve God is to seek to interpret wisely the events that do occur, not the ones that don't occur. The meanings on display in the natural world are multilayered, and they are not diminished by the fact that they take place within a tapestry of natural processes.

Among all the things that come about through natural processes, some show creative value, and some do not. Faithfulness to God includes learning to 'read' these processes in a way which is sympathetic both to their physical causes and structure (that is what science is for) and to their overall role and significance (that is what religious reflection is for).

When one gorilla extends a helping hand to another, it is genuinely help that is offered, and the fellow-feeling and helpfulness on display are aspects of the character of God, finding expression in this humble creature. When one gorilla attacks and kills another, it is genuinely a form of hatred that is offered, and the selfishness and violence on display are signs that truth and *shalom* is not yet fully present. Both behaviours take place by ordinary biological mechanisms; both also have the dignity of being exactly what they are: helpful or hateful. Commitment to God consists in seeing this, honestly and directly, and in committing to the project of embodying the helpful and creative attitudes in oneself, and promoting the true *shalom* throughout the natural world.

Let us now consider an example of moral judgment amongst humans, such as the judgement that murder is wrong. You sometimes hear people say that such a judgement is 'merely' an outcome of an evolutionary process. When someone speaks that way, they are being illogical in the following sense: it is the equivalent of saying that the strength of an arch is merely the outcome of the construction process of piling up stones. The point of the comparison is that, in fact, the arched shape genuinely is strong, as a shape, under compressive forces. The construction process is entirely irrelevant to this. The strength is an innate property of the sort of shape that an arch is. Similarly, murder genuinely is wrong. The evolutionary process does not alter that fact. The moral value is an innate property of the sort of shape in question—the embodied personal shape.

13.2 Better Metaphors

So far in this chapter we have explored one major aspect of evolutionary biology and explained why it does not and indeed cannot undermine the meaning that is on show in the natural world. In the rest of the chapter we will present ways of seeing, or getting the flavour of, what has been going on in the process of development of life on Earth. We need better metaphors than those which have been circulating up till now. We will return to the question of the pain in that process at the end.

13.2.1 Genes

None of us (the authors) is a professional biologist, but we do claim competence in understanding the main thrust of the big ideas in evolutionary biology. After thinking hard about this for a long time, we have come to the view that the dynamic of variation and selection that takes place in the Darwinian process is not best captured by the metaphor of 'selfishness' as in the phrase 'the selfish gene'. We can find other metaphors which also capture the idea of genes promoting their own propagation, but without the moral colouring associated with the concept of selfishness. One such metaphor is the metaphor of 'eagerness'. The idea is that the net effect of selective processes over many generations is as if the genes are eagerly asserting themselves, promoting their own reproduction without regard to others. This is reminiscent of the behaviour of children before they reach moral maturity.[1]

Words such as 'selfish' and 'eager' both carry anthropomorphic baggage, but sometimes one is willing to accept that cost in order to get the benefit of an improved intuition about how a physical process is working. The word 'eager' is better than 'selfish' simply because it is morally

---

[1] Andrew Steane, *Science and Humanity* (Oxford University Press, 2018).

neutral. We don't want to load our metaphors down with too much baggage. We want them as light and direct as possible. So we say that what is going on in evolutionary biology is, at one level, the reproduction and self-promotion not of selfish genes but of eager genes.

Of course, biological molecules are not really either selfish or eager, but the second metaphor is a slightly better one, because it is less loaded.

13.2.2 Making watches

The process of evolution is not altogether controlled. The genome of each type of organism has a very impressive degree of stability, but it can change a little from one generation to another, and the changes include a random component. At any given moment, it can and does go in all sorts of different directions, quite haphazardly. So this process of very gradual change moves around the space of possibility without, as far as we can tell, any mechanism or intelligent input steering it from one generation to the next. This lack of an overall director is what Dawkins alluded to by his metaphor of 'blind watchmaker'.

But an interesting thing happened when Richard Dawkins implemented a descent-with-modification process in software on his computer. It was a 64-kilobyte computer, and therefore a simple model, in which the notion of 'natural selection' was not very well implemented, but it gave some flavour of the process. After 29 generations it had started to produce two-dimensional static pictures or 'biomorphs' that looked a bit like bats, spiders, scorpions, tree frogs, and so on. The insects had eight rather than six legs, but Dawkins records his delight: 'even so! I still cannot conceal from you my feeling of exultation as I first watched these exquisite creatures emerging before my eyes.'

That feeling of exultation is, one may guess, a combination of three things. First, the satisfaction of getting a computer program to work for the purpose for which one wrote it. Secondly, enjoyment of the fact that the process of descent with modification and natural selection does what was intended. But third there is, in Dawkins' reaction, a sense of delighted witness, an enjoyment of the coming-into-being of something not directly created by the programmer, but for which the programmer provided simply the opportunity that something should come to be. It is reminiscent of the experience of parents as they witness the growing personal identities of their children, for whom, similarly, they have provided the opportunity for life and growth, but not completely dictated who they will be. Might such exultation be a pale, perhaps distorted but nevertheless authentic, glimpse of the joy of the creator of the world?

The physical mechanism of biological evolution—the long time-scale, the high degree of stability, with small amounts of random change, and natural selection—is, of course, a mechanism, an inanimate, unconscious process. It has no mind, nor mind's eye, and in this sense it is, obviously, sightless. But, to us, at least, it has a certain feeling of positivity and generosity about it. The openness is positive. It is the sort of process which does not dictate where things will go in detail, but gives an opportunity for things to explore, to find their own way into the space of possibility. It is an open-handed process, a generous watchmaker. Of course, it no more has hands than eyes, but this metaphor is an apt one.

Randomness is often portrayed as some sort of defect or problem. But randomness could also be named openness.[2] It is freedom from micromanagement. Is there, in fact, any difference between randomness and openness, in cases where the material has no conscious agency?

13.3 Pattern, improvisation, direction

But randomness is not the dominant principle in evolution. The petals of the daffodil, the muscles of the tiger, the antennae of the ant are not random, but show a great deal of structure. One might enquire, is this all just a massive fluke? Since, after all, evolution has been going on for a long time: there is plenty of time for random fluctuations to fall into chance patterns, like the appearance of a face in a coffee stain.

When one looks into it, one finds that the structure in biology is far from a fluke. The shape of living things is largely owing to profound harmony deep in the nature of the physical world. This is harmony of many types, including mathematical and physical and chemical and also extending into the patterns of social relations. The petals of a flower embody the same mathematical principles that allow arches and domes to function in our buildings; the muscles of the tiger exploit chemical and electrical signalling based on the beautiful and harmonious properties of the electromagnetic field; the antennae of the ant are shaped by basic mathematical scaling laws, and by aspects of engineering science

[2] For a positive appraisal, see Paul Ewart, 'The Necessity of Chance: Randomness, Purpose and the Sovereignty of God', *Science & Christian Belief* 21 (2009), 111–31.

concerning strength and vibrational modes of long slender structures. We could go on. The social lives of animals are also shaped by those truths that are relevant. The social existence of a pod of whales has notable commonalities with that of a family of elephants; in such high-level and complex aspects of the ecosystem there is largely richness not randomness.

Next, we consider some other aspects. The following is an extract from a poem[3] written as a meditation on one of the features of evolution that has been much remarked: the way in which solutions to the problems of living are improvised out of whatever material happens to be available. The anatomy of the giraffe offers a striking example:

> Up and down your splendid curve
> The long recurrent laryngeal nerve
> Winds its route the long way round,
> And, tells, perhaps, of your reserve:
> Slow to speak, long to think,
> Fragile as you bow and drink,
> Ungainly stooped upon the brink
> Of flickering thoughts you cannot link.
> Your family heirloom weaves its way,
> The tale of making-do at play:
> Jaipur to Agra by Mumbai?
> No shame at all! Your pride and joy
> To be a fellow-falterer,
> Holder-back and halterer,
> Trier-out and stutterer,
> Letter-out and alterer.

The recurrent laryngeal nerve is the nerve that runs down the neck in mammals, loops around the aorta near the heart, and then goes back up to the larynx. It is often used in discussions of Darwinian evolution as an illustration of the gradual adjustment or 'make do and mend' approach that this process involves. The circuitous route is arguably an inefficient way for this nerve to be routed—especially for a giraffe, in which it makes what might be called a detour by several metres. The evidence suggests that it came about this way simply because it is a workable solution that can be arrived at by the accumulation of small

---

[3] A. M. Steane, 'The Recurrent Laryngeal Nerve', unpublished.

changes, and because the evolutionary pay-off of re-routing it would not be worth the necessary disruption. Similar considerations apply to the wiring of the optic nerve in front of the retina in mammals and some other animals.

So this illustrates the Darwinian paradigm.

What the poem is doing is saying there is nothing wrong with that.

This inefficiency, this managing to make do, to live with limited apparatus, is a wonderful, beautiful thing. It is what life is all about.

The circuitous route may be part of the reason why giraffes don't vocalize much. They can make noises, but they are mostly silent creatures. People have used the giraffe as evidence that there is no design in nature. Other people have argued back that this feature is designed in for some good reason. Both are wrong. The design, or rather, to use a clearer word, intent, behind the world is that the world should be a world. A world in which fragile, awkward attempts at learning to live and to love can be made.

Let us say it again: managing to make something wonderful out of whatever bits and pieces happen to be to hand is, truly, madly, deeply, what life is all about.

The poem muses on this a little, and also asks for a more general slowing down, a taking another look and being less quick to judge:

> If someone said, 'let's make a snip,
> Re-route that branch-line, make it quick,'
> I'm not sure that I'd vote for it.
> At least, I'd first stand back and look;
> Look, and watch, still and long.
> My thoughts would take an arctic turn
> And hear the humpback whales' song
> And run the race of marathon
> And pause to stop and stay, and weep
> At trees floating in the creek,
> And factory fires and mission-creep,
> And oil welling in the deep.

Because the mindset that is quick to talk about efficiency is, too often, the mindset that cannot see the true cost and value of things. It has led to people working under appalling conditions for the benefit of a few rich business-owners and shareholders. Too often, also, it has turned into the mindset that leads to war and environmental devastation.

Let's enjoy and celebrate the fact that biological evolution shows much signs of improvisation, of making do, of finding new uses for old tools. It is often a story of making something wonderful despite the imperfection of the component parts. This is very much like creativity in human art and artfulness. It is part of the practice of *up-cycling* and the reason why Wallace and Gromit are such fun. It is also a law of spiritual life, of how we must learn to live. Here is something of the character of creative love, and so of God.

In recent years people have often attributed the wrong meaning to the word 'design' when it is applied to the natural world. The word 'design' can mean both 'intention' and 'detailed prescription'. There is no need to assume that every minute detail of living things was pre-decided by a divine plan. We have no way of knowing for sure, but the world looks very much as though its originator has allowed it a great deal of freedom to develop in its own way. As far as we can tell, that is the intention. And that seems to be in keeping with the dance-like quality of the Kingdom of God as Jesus described it in so many enigmatic hints and pointers.

The net result of this development is not completely unconstrained, because it is the result of exploring the space of possibility not impossibility, as we already explained. Darwinian evolution has discovered eyes and wings, livers and legs, because these things are possible and help living things to live. It has not discovered a faster-than-light propulsion system, because no such system is possible. And it turns out that the space of solutions to the problem of 'how to generate more copies of these genes' is not a hotchpotch, but has its own structure. Very many characteristics of living things have been repeatedly 'discovered' by this process. The generous watchmaker is a process very likely, and perhaps guaranteed, to produce increasing degrees of complexity and increasing degrees of awareness in communities of living creatures.

The aim of these reflections is not to impose on this story metaphors which it won't bear, nor to see it all through rose-tinted spectacles. The aim is to be true to the story, to see it truthfully, and help others so to see it.

13.4 Born into pain and joy

An abiding difficulty in the facts about natural processes, for anyone who trusts in God, is the presence of so much pain. And many animals have only a brief experience of life, far short of what one feels they might legitimately have expected to enjoy or that we might expect on their behalf. After investing in the creative process of bringing animals into being, it strikes a sensitive person as wrong to allow them to die young, and wrong that there should be lingering, drawn-out suffering. This pain has been there throughout millions of years, and pain is experienced by every creature that can experience pain. This pain and death is not caused by human wickedness, but is an integral part of the processes by which the ecosphere has developed. Humans, after their relatively recent arrival in this process, have certainly experienced pain, and they have exacerbated the problem by their own choices, but they are not the root cause of parasitism, predation, disease, and death.

In some parts of the Christian community there has, historically, existed a belief, based on a fond hope and a literal reading of an isolated story in the Bible, that the natural world was pain-free before the arrival of humans, and that human life itself could have been free of difficulty if only people had made better choices. This intuition is nowadays unconvincing, and indeed it is hard to imagine how it was maintained in the first place. Without death the planet would very quickly have become overpopulated. Presumably people had in mind that some miraculous process would prevent this, but it is hard to make sense of the idea, and even harder to find it in the Bible. As we saw in Section 5.5, Genesis 1 portrays God appraising the structure of the physical world as *tôb*, fit for purpose. God is not declaring that the created world is a sort of idyll, totally free from pain. God is declaring that the physical order is well made and has the combination of properties that it needs.

The connotations of the word *tôb* are not about any idyll but about something more pragmatic. A musical instrument-maker will make sure that the keys or strings of the instrument function correctly; they may then pronounce the instrument 'sound' or 'good' in this sense. Similarly, a builder who made a house might push against one of the oak beams that supports the roof in order to assess its strength and firmness, and he may pronounce his workmanship sound or good. This is the sense of the text in Genesis chapter 1. The natural world is sound because it works. And this is why science also works. This soundness of the world is the fundamental reason why scientific study is appropriate and productive.

There is no indication in the Garden of Eden story that an easy life would have been possible in this world if people had made good choices.

The story presents a picture in which work, not ease, will be required. It does not say that life first became painful after people disobeyed God. It says that life became more painful. The point we are making here does not depend on the literary genre of Genesis chapters 2 and 3. Whatever is the amount of symbolism in this text, the account describes a world which, from the outset, will require the contribution of human work 'to till the ground' and in which pain is part of the experience. We don't doubt that such work can at times be delightful, but sometimes the sense of joy and worth comes only after a period of exertion, which is the necessary route to the goal. Moral failure comes in when our efforts are out of tune with God and with our own better nature and that of the world around us; such efforts are, sooner or later, counterproductive.

We shall not attempt a full exegesis of Genesis 2 and 3 here (though we have touched on it in Section 5.5). The main point is that, as with all of the Bible, we have a duty to try to find out what is the thought-space of the writers in order to understand what they are trying to tell us. 'If we don't do that we are not really listening to the Bible, but simply hearing our own voice echoing off its pages.'[4] In the Bible the word 'death' often means something other than physical death. 'In the Bible, death is the reverse of life—it is not the reverse of existence…It is a diminished existence, but nevertheless an existence.'[5] That is the sense in which death entered the world through sin: it was a diminished existence, a spiritual death.

One may try to imagine what communal human life would be like in the absence of closed-heartedness and ill-will. If it could have happened, then surely such human life would have been challenging and sometimes very painful, but equally rewarding and joyful, and overall, immensely worth living. Its pain would at least have been free from guilt (except for the false guilt that accompanies some mental afflictions, and perhaps those might be avoided in a good society). The extra pain that sin introduces is enormous, totally changing the whole structure of human society, and putting many non-human animals into further suffering too. Nevertheless, with or without human folly, life has finished in death for all living things, and animal life included predation and disease long before humans existed.

---

[4] E. C. Lucas *et al.*, 'The Bible, Science and Human Origins', *Science & Christian Belief* 28 (2016), 74–99.

[5] H. Blocher, *In the Beginning* (InterVarsity Press, 1984), 171.

We conclude, then, that pain and physical death are part of the God-given pattern of life on Earth, whereas spiritual death is a breakdown of that pattern. And the former part of this, the pain that is the result of processes for which humans are not responsible, remains shocking. If, as we think, it is right to ascribe moral responsibility to God, then God bears a heavy responsibility for having brought into being, and sustained in being, a world in which life-shattering processes go on for no fault of the creatures who bear the brunt of what results. A follower of Jesus should not, in response to this, try to construct arguments which suggest that such things are 'good' really, in some hidden way, or can be justified by some good outcomes that may result. The attempts which have been made to do that strike us as unconvincing and lacking in the sort of wisdom which accompanies compassion. Rather, the Christian view should be that, through a process of which we only have some vague inkling, God bears the responsibility in a way that can, ultimately, merit our faith in Him, and this includes a bearing of pain, a fellowship with the insulted and the humiliated.

This will not address the spiritual needs of someone in a present experience of severe pain. We are emphatically not engaged in a heartless exercise of trying to address the needs of such a person merely by subtle arguments in a book. But we do wish to acknowledge the genuine questions which arise, if only to admit that the problem of pain is not solved by an abstract calculation; it can only be met in other ways.

God is not somewhere else, watching, but right here, being. Faith in God does not mean that one can arrive at a justification of all pain, or of lives cut short, and it certainly does not allow any form of argument that tries to justify the pain of others in order to simplify one's own existential struggles. Faith in God means that our sense of objection to injustice is itself God-like.

Nor is it appropriate to try to form schemes of calculation of total pain, or to try to weigh it against total joy, if such schemes are merely attempts to judge the world, to declare it finally either bad or good. The only worthy role of such surveys is as an aid to relieve suffering through understanding it better. In short, Christians ought to resist the temptation to construct edifices of argument whose intention is to make us contemplate the pain of the world with equanimity. Atheists ought equally to resist the temptation to exploit the pain of animals as a support for the promotion of atheism. It is the exploitation that is problematic here; the idea of using someone else's struggle not as a prompt to your efforts to help, but as a way to bolster the case for the rightness of your own opinions. Also, when remarking on the pain of the animal world, for balance one owes to one's audience a frank admission that scientific study of biology and ecology is mostly reported to be a very positive experience.

If one claims that the physical world is objectionable in some way, for example because it involves pain and randomness, then one is already asserting some standard whereby things can be deemed objectionable. Hence, the very impulse which asserts 'this is meaningless, atheism must be right' is itself an impulse which wrestles with meaninglessness and tries to force it into a good role: the role of teacher. But one can also assert, 'I refuse that this be allowed to have meaningless outcomes' and go on to respond in other ways. These two reactions were explored with great insight by Dostoyevsky in his novels, especially *The Brothers Karamazov*.

Fyodor Dostoyevsky himself experienced epileptic seizures, poverty, and imprisonment in Siberia, as well as such psychologically abusive acts as a mock execution. In prison he witnessed the worst forms of brutality, including towards children. His novels do not hold back from contemplating the horrible suffering that children as well as others have had to undergo. Dostoyevsky's Russian Orthodox religious faith was never at ease, always in testing dialogue with his own scepticism, but it was important to him that this faith was the voice of the common people, as opposed to the ruling class or the intelligentsia, and to him the figure of Christ was immensely attractive, as one who both understood the bitterness of human life and offered a genuine hope of transcendence. *The Brothers Karamazov* is both a passionate and a philosophical novel, which enters deeply into theological and ethical debates. We cannot summarize the book here, but it is by such deep and honest work that the problem of pain is to be addressed, in so far as it can be addressed by writing.

His own experiences led Dostoyevsky neither to pride nor to despair, but to appreciate the very process of life as an incomparable gift. This is the same refusal as the one we already mentioned—the refusal to say the world is fundamentally absurd. This does not mean saying that every last detail has been deliberately introduced or caused. To repeat, it is unhelpful and wrong to say that every instance of suffering is 'for your good' in some unfathomable way; rather, we want to say that our response can be good. Our response is not absurd.

One can understand a legitimate role for expressions of anger and dismay at the pain of the natural world. These are proper and moral if they are genuine and motivate effort to make things better. But not all expressions of anger and dismay ring true; sometimes they are a tactic in a dispute—for example, a dispute about whether trust in God even means anything at all. But the only honourable response to the pain of others is not to exploit it for one's own ends, but to share it and seek to overcome it.

13.5 Concluding reflections

In this chapter's final section we will present some reflections of a more personal or subjective nature.

Well on in the course of writing this book, we (the authors) had a brief discussion in which we found it natural to exchange, informally, some impressions of how biological discoveries strike us in the round, and what we find comfortable or welcome, what difficult or unwelcome.

One thing that biology, especially genetics, has made more and more apparent in the past fifty years is the degree to which there is continuity between humans and the rest of the natural world. We find this rather lovely and welcome. When we reflect on the experience of an animal close to us in the evolutionary family tree, such as an ape, it is natural to have some sense that their experience of life is not utterly alien to ours; there are signs of both worry and delight there. The beginnings of empathy and morality are there in the lives of the great apes.

This connection between humans and other animals has sometimes been interpreted so as to diminish human value. It has been used to imply, for example, that our efforts at empathy are just instinctive reactions programmed into us, with no more meaning than that of one worker ant smelling another. What such interpretations fail to realize is that the argument works both ways. If we are like ants smelling one another, then ants are like humans forming impressions of one another. The second way of stating the connection is not unscientific or unjustified. It is just as scientific as the first, as long as we make the effort to understand the difference as well as the similarity which is involved in such comparisons. In short, the connection of humanity to the rest of life on Earth does not diminish humanity, but rather heightens our regard for the whole natural order and the possibilities it contains. A similar conclusion was expressed long ago by Francis of Assisi.

Andrew Steane recounts:

There is, in the contemporary world, a widespread sense of tension between Christian belief and what we learn from biology. What can fairly be said about biological processes? Are they, in all honesty, the sort of death-knell to the Christian movement which they have been said to be by some? Is the lesson, 'look, the world is not the outcome of any deliberate act as people once thought, and the movement you joined will fade away because it cannot deal fairly and honestly with well-established data'?

My experience has been one of accepting the raw scientific facts about evolution, but being sceptical of the narratives attached to them. I have carefully plodded a journey away from the spirit of the age, with a view to finding out whether biology can be seen in such a way that it does not diminish the values celebrated in the Universal Declaration of Human Rights, for example, and more generally does not contradict the high-level truths on show in the art, literature, and political freedoms of the human race, and in the lives of ordinary people, by suggesting that they 'are really' something else altogether, such as subtle reproductive strategies. In short, I have sought an understanding of science which is not oppressive to the human spirit. Where I have ended up is a better place, I think, but not altogether comfortable.

By an oppressive notion I mean, for example, the notion that the beauty and grandeur of the natural world is a misleading appearance, and it 'is really' a largely brutal place in which fast-moving individuals with 'an eye to the main chance' get ahead by exploiting their fellows. I mean the sort of mind which, when viewing a million gulls on a cliff living alongside one another almost entirely without violence, chooses to focus attention primarily on the few which are engaged in a dispute, or on the act of stealthy violence when one gull gobbles up the egg of another. When I was first introduced to the latter kind of fact, it was presented in a way which tried to sell the idea that here is 'the truth' of it; this is what all the gulls 'really want to do' and would do if they got half a chance. I found this sort of sales pitch to be oppressive. In fact, what the gulls really want to do, in so far as they have some rudimentary will, is broadly what they do do: live alongside one another in a mostly trustworthy manner. That is what I now think.

It will be said that the gull which exploited its neighbour really did 'get ahead' here—got a free meal, and consequently a heightened chance of passing on the genes which predisposed it to do that. But the point is, does this act of parasitism constitute 'getting ahead'? Is that what we think 'ahead' means: exploitation and self-promotion? And is that the future of the gull population: selfishness and war? Actually, no. The egg-thief displayed a behaviour that cannot ultimately dominate, and this is a beautiful truth.

It is true that, in any given second, some animal somewhere on Earth is suffering. But this is, to a large extent, simply because there are so many animals. It is not because animal life is 'beyond all decent contemplation'.[6] Life makes possible the experience of both pain and joy, and there is every suggestion of joy and adventure in the lives of animals with sophisticated nervous systems. Their existence does not have the appearance of being dominated by fear, for example. In a herd of herbivores on an African plain, most spend large amounts of time grazing peacefully or walking calmly. They are always on the alert in some measure, but not in fear. This is not to deny the fear and exhaustion of a chase to the death, but to put it in a correct perspective. In fact, many scientific researchers and others are drawn to long study and contemplation of the lives of animals, and it is rare that such people are simply repelled by what they find. The picture has elements of both beauty and ugliness, but it is often found overall to be positive and fulfilling.

It used to be the case that television presentations of natural history showed an almost exclusive fascination with predators and predation. Thankfully, this is less so now. We are now, more and more, getting the chance to absorb a more balanced picture, in which predation plays its part, but is not afforded quite so much exclusive air-time. We are being invited to admire the patience of the wildebeest as well as the strength of the lion. We can contemplate the protective efforts of the mother whale as well as the cunning strategy of the killer whale.

In this latter example, killer whales demonstrate their intelligence and determination when they attack as a pack. They also demonstrate a near-complete inability to recognize in their thoughts, or make allowance in their actions for, the value and pain and legitimate hopes of their prey. In viewing this we can admit the value of intelligence and determination but this does not mean we have to say that pure aggression with no thought for the struggles of another is also good. Neither does it mean we must say that there is no such thing as either good or bad, right or wrong, value or its opposite. Rather, we are allowed to exercise our precious gift, as humans, to see and name where the value lies. We don't have to sell our birthright and announce that there is no value or disvalue but only machines and mechanisms going through their motions.

Let us comment a little further on this point.

[6] Richard Dawkins, *River out of Eden* (Basic Books, 1995).

There has been a movement in Western culture, building over a long period, towards the view that the natural world is entirely neutral in terms of value in and of itself, and human freedom includes, or even consists in, a freedom to come to our own subjective opinions about what is good or better, and what is less good or not good. The idea, apparently welcome to many, is that to accept this is the way to be correct in handling the intellectual issues, correct about science, modern and grown-up. In short, there are no absolute values about what goes on in nature, but we may have subjective opinions.

We (the authors) feel that something has gone astray here. This notion that value is purely subjective does not seem to us to constitute intelligence or wisdom or freedom. It is rather the opposite, because an intelligent reflection realizes that one cannot deduce this as a conclusion of any argument, and even if one tried to, someone may always reply that the argument was without merit because it is without value. Therefore, it is hard to see how the claim that value is purely subjective merits to be called intelligent, though it might be someone's opinion. We don't think it is wise, either. This is partly because it is so disheartening. It is just about the most disheartening thing anyone could say. Our whole inclination is to rejoice when a penguin takes so much trouble to care for its egg or feed its young, and to be dismayed when a parasite eats its way into an eye. We don't want some modern-day pontificator coming along and telling us we are merely being emotional when we make such judgements, and we could equally see it the other way around if we wish. If this is just subjective, then we are diminished.

Perhaps the modern hesitation about absolute value is really a way of saying we cannot ever be sure about our abilities in this area. We (the authors) would like to affirm that what humans have is, genuinely, a sensitivity to absolute value, but our sensitivity is never perfect. What this means in practice is that we think the remarks we just made about penguins and parasites are not completely off-beam, but we can never be sure that we have read the book of nature correctly in all respects.

Where we three imperfect commentators finally arrive, after our plodding journey, is a combination of joy and grief. We love science, including genetics and evolutionary processes, and we love the marvellous display of wonderful life all around us, and the far-off creatures shown to us by fascinated investigators and highly skilled camera crews and carefully argued scientific papers. On the other hand, we also go about mourning and grieving for the pain of the world. The whole world is in travail, yearning for the peace it does not have. It flames into life, and it is also crushed into death, and this is where we find our work to do.

When we discussed the subject of quantum physics in Chapters 7 and 8, there was a sense of consonance with the life of faith in God. To someone willing to say that the world gestures towards God, it seems fitting that the world turns out to be, at its fundamental physical level, somewhat open and mysterious and not completely captured by any simple physical picture. When we contemplate the process of variation and natural selection, on the other hand, there are moments of dissonance. It is not so easy to know what we should do with the information, or how we should react. We are left in a position of not knowing, only able to offer ourselves, by way of support, the intuition that meaningless pain will not have the last word; the precious moments of fun or charm or courage or sharing are the fusion points of this world, the last that shall be first.

14

A conversation about naturalism

This is the transcript of a discussion between the authors which took place in Oxford in May 2017. It is based on the concept of naturalism in philosophy, for which the Oxford English Dictionary gives the following definition:

> The idea or belief that only natural (as opposed to supernatural or spiritual) laws and forces operate in the world; (occasionally) the idea or belief that nothing exists beyond the natural world.

HH: Do you consider yourself a naturalist?

AB: I think that the natural world, what I would call the material world, has some reality to it. It's the world in which I live my life, and in my professional work it's the world that I study. But this doesn't mean that I think that this world is all. Indeed, I would say that this world is just a part of reality, albeit it's the material world in which we live out our life and our faith and our spiritual existence. For me, that gives importance and a kind of dignity to the material world.

AS: To me, the word 'naturalism' is a word which stands for a certain perspective on things and if one defines the word 'naturalism' in the way I normally understand it, then no, that's not me. That's not my philosophical position. There is the world of sight and sound and all that which is embodied by the gravitational and electronic and quark fields that physics describes. If naturalism is the view that that world constitutes the whole of what can be considered to be real, then no, I don't think that would be true. However, I think what naturalism is also trying to affirm is something I do subscribe to. I think that people who are drawn to naturalism are saying they feel that they want to affirm a very strong sense that it's here, now, in the reality of our molecules and bones and flesh that we exist and interact and live out our lives, and that's valuable—that material existence is itself the valuable thing. That I agree with. I don't think that this captures the whole truth, but it's precious. We are not at all invited to distance ourselves from the material embodiment of who we are.

HH: From the traditional philosophical categories, some people would judge, from what you have just said, you sound like you are a dualist, so you believe there is physical stuff and there's non-physical stuff. Are you happy with that formulation?

AS: Not really!

AB: I think there is difficulty with terms here. I'm a researcher whose home in Oxford is in the Department of Materials—my scientific home, I mean.

HH: You're a 'materialist' by trade!

AB: If you are in the Department of Physics, you are physicist, if you are in the Department of Chemistry, you're a chemist, so I ought to be a 'materialist'. However, the word can't get away from its history and a lot of people would think that the word 'materialist' means someone who thinks that this material world is all that there is; and I think that's profoundly wrong; I think it is profoundly inadequate and I'm equally nervous of saying that there is a kind of material stuff and then there's a non-material stuff. That doesn't work. The word 'stuff' just, of course, means materials in German or in Old English.

AS: ...and that's what I think it means as well. You can't say 'more stuff' because then it's just more stuff...

AB: ...materials means stuff. So I don't want to say there's material stuff then there's non-material stuff, I just think that's just nonsense. I think the words I am using help to reveal what nonsense it is, but nor am I content to say that the material world is all that there is. Yet I do acknowledge the material world is the time and space within which we are ourselves and we exist and we do things and we relate to other people through the material world. I'm relating to you now through the acoustic waves that are mediated through the air.

AS: Two thoughts...in reaction to the question, 'so are you a dualist then?'...I find that easy to react to, and say, no I am not. What 'dualist' seems to stand for, in the way that it's normally used, I don't find helpful. So if I asked myself 'how could I put what makes sense to me in a more affirmative way' I think I would do it by virtue of some analogies, metaphors, things like that. So I asked myself the question about music—would I say that music is real? I am not a professional philosopher but I know it's the sort of question which philosophers think about and they come to agreement on what an answer to a question such as this might be or might mean. I think the word 'real' for me does include that 'music' would be real and I don't just mean sound vibrations of course. I mean that which the sound vibrations are carrying or are conveying—that's somewhat more abstract. You can't necessarily put your finger on it directly and yet neither is it completely removed from the material world. So there is a sense to me in which there is a reality to music and not all sounds are music— some are and some aren't. So that helps me a little as a way of getting at what I want to say.

AB: Would it be helpful to introduce a technical distinction between method—how I do research in my laboratory—and ontology—how I think about the essence of existence? In the laboratory I behave as a methodological naturalist. Everything that I try to study and measure I think of in utterly materialistic terms, and I think in terms of temperature and magnetic field and voltage and current and frequency in internationally defined units. I thus methodologically reduce everything to naturalism. But then in my life as a whole, I realize that there are wider aspects of being alive, if you like of existing, that cannot be reduced to naturalism alone. For that reason I am not an ontological naturalist, even though I recognize that our existence is indeed lived out in the natural world of material stuff.

AS: I would like to take that in two directions. The first direction is just throwing in another analogy. Another analogy which is somewhat helpful, I believe, is the analogy between a text and its meaning. So when we read a book, we are absolutely confident at the outset that, if we are dealing with any sort of ordinary book, there's a series of black and white symbols on paper but we know that it means something and there is some sort of shared language with the author and we are expecting to receive that meaning. I think that there's a sense in which the natural world is also a carrier of meaning and that's really quite a strong statement. That gets me a long way to expressing what I mean by saying naturalism isn't quite enough to capture the truth of things. The truth of things is that the natural world does carry, I think, a meaning over and above simply equations which physicists discover and things of that kind. So the other thought I wanted to add, reflecting back on both of those, is this. It's not an unusual experience when listening to music that people sometimes feel that the music conveys to them something rather personal. People can feel an experience of mercy from music?! I think that's genuine and there is some sort of truth-speaking that is going on in a case like that. There is a sense in which something is being merciful to us. I say 'something' but of course I think really it is 'someone'. After all, how can a 'thing' be merciful? That is just the beginnings of a feeling towards how I would put a more positive statement, a statement of what I do think. But I'll leave it there and ask the question to Hans.

HH: I agree with the spirit of this. In particular, I agree that we are caught in a bit of a dilemma because, on the one hand, none of the three of us want to say that the 'material world' is all there is—we are not inclined to say that—and yet we are also inclined to say, 'there's no other 'stuff'. Again, note how language gets us tied up there, because often when we ask about 'Does something else exist besides the things we see?', the natural thing for us to do is to project qualities of the things we see, the things that we live around, on to the 'other', the other 'that which exists'. This 'other' being whatever it is. I would say that there is no other stuff than the 'stuff' we live in because the other is not 'stuff'...like the stuff we live in. The language we use here gets us caught in a bit of a tangle. However, I still feel that, in my mind, there is another question that has to be answered. It is that the analogy to musical notes, or let's put it more concretely in terms of physics—sound waves and then music—is not adequate. That analogy does give us some sense in which there really is something there that really does transcend the physical sound waves, and the same goes for the letters on a page and the meaning of the written text, it's something that really goes beyond it. However, I think that analogy also has limits, because if we are talking about God, then it doesn't fully work. Some people would say 'yes, that's a good analogy' for the way that God is to the natural physical world, that somehow God is the meaning behind the natural world in some sense. But if you take away the sound waves, the music is not there anymore; however, I don't think that I would be inclined to say that if you take away the material world as we know it, God wouldn't be there anymore, because I think the dependency relation in that case is reversed. I think that the material world depends for its existence on God. But I don't think we would be inclined to say that sound waves depend for their existence on music. I might be inclined to say the music depends for its existence on sound waves.

AS: I understand what you are affirming—I wouldn't say that last statement actually, but I agree with what I think is the gist of what I heard you say...

AB: I am sure the music can be there without the sound waves—sure it can! Surely, we have all had this experience . . . 'I can't get that tune out of my head'. The music is there. The sound waves aren't—but the music is there.

AS: And any well-trained musician can almost 'hear' the music just from reading it on the transcript/page. So there is a sense in which the music is being conveyed to them through reading it as a written script, which is interesting. But nevertheless I agree that none of my analogies works completely satisfactorily. I think that surely it must be that the world depends on God and not the other way around and also that the world doesn't exhaust all that God is and also that much that goes on in the world is not what God would wish. So this world has this independence from God, while nevertheless only being able to be there because God decides that it will continue.

HH: I think this point is well taken. I think there's a subtlety. Obviously, one can say, if you are playing the radio and you blast off in a rocket and you pass outside the earth's atmosphere, in empty space you can't hear anything anymore, without the atmosphere to carry the sound waves—there is no more music, in one sense but in another sense 'music' will always still exist. Take some famous symphony, e.g. by Beethoven. It would still exist even if no one ever played it ever again! It would still exist as a 'thing'. There is that symphony that was created—it will exist forever in some sense. So I think, in a way, that it does draw the analogy a bit more closely, although we still don't assign some agency to these abstract entities, symphonies, things like that.

AS: And they are created things, not Creativity, or that One who makes our creativity possible.

I think it might be helpful to mention a way of thinking that we find in the Bible, where there is quite a strong connection between 'truth' and 'action'. There is a notion of truth as something not completely passive, but actively accomplishing things; also truth as something that one should not only understand but also obey. Another example is the idea of 'logos' which is translated as 'word', but as I understand it, this also has somewhat of the quality of a verb, or a truth which is intimately connected to the shaping of the world, a bit like what we call laws of nature, but more personal, or coming at it from the other end, so to speak. And this also reminds me of the phrase 'God is spirit' that Jesus used when talking to the Samaritan woman in John Chapter 4. That was just part of a conversation, of course, and should not be taken to be any sort of definition, but I mention it because it connects to the things we have been saying.

AB: To move beyond the music analogy to the description of God becoming man (and inevitably, anybody trying to describe this is going to struggle to describe it, and they have struggled)...John, in his Gospel, took this integrated concept of 'word' and 'communication' and 'underlying principle of the universe' and so on, and said that that became physically embodied in Jesus. Embodied in a way that is more profound and more integrated than had ever happened before. God had become material in that event and in that person.

HH: Yes, that is certainly true, although we think that in God as a trinity, God the father would have pre-existed, did pre-exist that event, so it's not that the incarnation of Jesus somehow was God coming into existence in some sense.

AB: God expressing God's own nature in physical ways wasn't something that had never happened before. The Hebrew scriptures struggle with how to describe lots and lots of different ways in which people experienced God, before the time of Jesus. Sometimes in terms of a quiet voice, sometimes in terms of apparently a physical presence in human form and other different ways too. I think they struggled to describe it in terms that were familiar or perhaps unfamiliar concepts then, and we would have to work hard to think: how do we now understand it, in terms of the world view that we live in, which includes a world view in which science has proved so spectacularly successful?

HH: Do you think that our current knowledge of science allows us to understand how things like that are possible? E.g. God spoke to the ancient Hebrews. Do we have to now think, how did that happen? Think about someone like Rudolf Bultmann, who said that you can't possibly believe that's literally true, given that now we believe that any communication occurs through a physical medium. For me to say something, for Andrew Steane to hear me, there have to be sound waves that transmit from me to him. If we weren't connected spatially or temporally—there's no way I could speak to him. How does it make sense to talk about God speaking to us, now that we know speech, communication requires physical substantiation?

AB: I wouldn't want to put anything off limits to science, I wouldn't want to say there are aspects of existence where you are forbidden to bring your scientific thinking to bear. But it may not be terribly useful.

It may not be very fruitful to bring that sort of enquiry to bear on this kind of issue.

AS: I am not sure about that. I think that there is sometimes a sense that some behaviours are appropriate and some are not. And the enquiry called 'scientific' depends on what you mean by that. I'd hate to do it, but to make the point I will introduce an extreme example. An extreme example would be to carry out an experiment on someone to determine how much pain they can resist, without their consent. Would that be called a scientific study?

AB: Well they did it!—It was done by the Nazi doctors in World War II. In that case it was how much cold their victims could bear.

AS: I didn't want to spell that out, but if that is within the meaning of what we mean by 'scientific' then, no, it's not always appropriate. That of course is an extreme example but then you become aware of other cases where it might be an example of something more positive, in which God made himself known to someone in rather a remarkable way, and then we start asking—so was the air vibrating when Moses stood before the burning bush? Is that the appropriate question to ask? There is a sense of holy ground there. And I would say, 'Go away with your stupid science questions!'

HH: I agree and I also think it would be easy to fall into a related temptation. Imagine yourself a young eager scientist, learning the tools of the trade and thinking that 'I can apply these tools to everything.' If you have a hammer, everything looks like a nail. 'I have electromagnetism now…I am going to understand how God spoke to Moses.' But there is also, in principle, a reason to think there is going to be a point at which, in such an analysis, it will break down because we think that, insofar as these laws are true, they are true of things that God created, but they wouldn't be true of God because God isn't governed by the very structures that God instantiated in creation. But anything we have in science is intended to describe one of these structures instantiated in creation. So if God breaks through into creation, he communicates to us in some way. In that case, there is going to be something that just doesn't fit into the description of natural processes.

AS: That's right, but I would also add that the experience of *encounter with God* need not be miraculous. The analysis in terms of electromagnetism and things like that might fail simply because it is not adequate to grappling with the larger meaning of whatever is going on in any given case. Indeed, as many commentators have emphasized, if the story of Moses before the burning bush were a form of story told long after the event and shaped by the teller, then that would not change its impact and importance very much, because what is important is the things learned about the character of God, and the type of leadership subsequently shown by Moses. I find this part of the account speaks powerfully of a genuine encounter with God. But I also think that it is correct, as a matter of intellectual duty, to acknowledge that the story of the Exodus is not simply a literal history in the way it has been thought to be by many. I mean, the archaeological and historical evidence indicate that that is not the sort of document we have inherited, and the internal evidence is consistent with that. But none of this changes the fact that it is a genuine example of an interaction in which God was one of the parties.

Traditionally God's interaction with the world has been seen as taking various forms, among which would be included Creation and also Providence, inspiration, sanctification, and miracles—it's roughly that grouping. Or perhaps one could put them in other terms? What you have just described is somewhere in the vicinity of inspiration and miracle, and a bit of sanctification—so we feel that it is God at work. I think what you are saying is the following. Suppose one had a scientific analysis of what goes on in a person when they become more as they should be. You are not sure if you would say that such an analysis captures the whole truth of what's happened and I think that I am in agreement there.

HH: I agree with only a slight addition. I think all three of us have agreed that there is this pragmatic, 'Is it the right question to ask?' And we all agreed, 'No, you have missed the point.' So if you go to the burning bush example and you say, 'Let's think of the physical mechanism,' then you've missed the point of the story. But there is a further issue regarding 'Why can't we just go ahead and wield the scientific tools everywhere?' It is because we think as a starting point (I don't want to say 'a priori' because it has a bad connotation) in the way we write elsewhere in the book. We are *coming from* a position of faith, we are not expecting a purely scientific interaction with the natural world to lead us to a position of faith. When we come to the natural world, we are coming 'with' faith; this is God's creation and manifests God's glory and we believe the things we've been told. I mean, told through careful thought and remarkable action, and so on, but I won't get into that. We consider it to be true and important that Jesus is God's Son and that God has communicated through history. In consequence, we expect that science is of great utility in a limited domain and then at some points there are attitudes that will just cease to be appropriate. But we don't really have a principled guide for 'Here's a case where we can apply it... here's a case when we can't.'

We were saying about 'sanctification'....I actually think that's the primary, one of the more profound examples of where it's still very appropriate to use the language of divine action in the world. Actually, I think much more appropriate than physical wonders. So if you say something like 'I was lost, but now I am found'—how did that happen? A psychologist could do a psychological study: What was the environment you grew up in? What did you eat for breakfast, the morning that you had that experience? But I think in certain cases we would say that that's missed the point for us. No matter how tightly you analyse that, we believe that's a case where God interfaces with nature.

AS: I have another example, just to put that same point in another way, which I find makes sense to me. One person asks, 'Is God real? Is God not real?' and they want to analyse that philosophically and OK, there is some place for that... and their friend replies, 'I don't know, but since looking to God I have overcome my addiction, I am a better parent—things are going better, I think I will stick with it.' So, that's another way of saying the same thing. Also, I read of advice given by a father to his adult daughter as he was facing the possible near end of his life—he had cancer and they spent many precious moments together, and among the things he said to his daughter was a sense in which he wanted to say to her, 'Trust the Universe.' He meant to say, 'Go forward in your life, don't worry too much. Life is good, I know I am dying of cancer but nevertheless there's a sense in which you can trust yourself to the wider world.' He used that phrase 'trust the Universe', because he had not found, for him, a way in which he finally wanted to use the word 'God' for that. Now the meaning of the word 'trust', I think, is to do with persons and there is a sense in which you can't actually trust an impersonal universe. At least the word 'trust' captures something larger than that. What that father was trying to affirm, I think, was either the same as, or is very close to, what we want to say when we say 'trust God'. But people are so confused about God, or find the word so overlaid with odd assumptions, that the idea of 'trust the Universe' may often be more helpful than the muddled message that is conveyed when people say, 'trust God'. If I were in conversation with that parent, on a day when he wanted to have the conversation, I would put it to him that what he wanted to bequeath to his daughter was an attitude that blossoms into 'trust God' as we get a better vision of God, but I admit that many people have never found a way to see how that is actually a good way to say it.

HH: Yes, but I want to develop that and then turn it into a question as well because I wonder why are people (not the real question yet) resistant to making that extra move. People in this day and age especially. Of course, it is well known that there are statistics saying people are more spiritual than previous generations but less religious. It's especially true in the United States, and somewhat true for the UK. People say, 'I am spiritual and believe in the transcendent, I am not willing to say "that's God" and I'm not willing to be involved in the church. I'm certainly not willing to go as far as saying Jesus was God's Son.' I think sometimes there is sort of a good reason. People have their reasons and what I want to ask them is sort of tangentially related but still related to the question of naturalism. I think if we extend some charity to people we disagree with in other ways, to people who find it important to say, 'I'm a naturalist', and who proclaim this, I think we're in a position to see that there is something right in what they are saying. I want to ask what you both, coming to it as scientists, and therefore sorts of naturalists as you said earlier, would say to this. I mean, what do they get right? People who say, 'I espouse naturalism', or 'I'm a naturalist'; part of it may be that they want to distance themselves from God, but part of it may be that they want to capture something good that historically people of religious faith have often not expressed or recognized. What is that or what are those things? What are the good things about naturalism that attract people?

AS: I can think of two things: first, a combination of real honesty about evidence, a great sincerity and honesty about the desire to be careful about evidence and reasoning, that's one aspect that I can think of which is good and right. Another one is just a simple affirmation of the material world, which I think is also a good thing. And within the Christian community there are people who are happy to affirm the material world. Francis of Assisi is a famous example, but I think he is just the figurehead that represents a wider phenomenon. Personally, when I read the Gospels, I get the feeling it is Jesus' position as well. Francis of Assisi is expressing an aspect of what I also see in Jesus, in his willingness to let the world be a good embodiment of a way of being a person.

AB: I think that's right. I think he talks an awful lot about money and there are various ways of thinking what money does, but one of the things it does is enact a way of distributing and controlling limited resources. This may be limited material resources, like fields and houses, or it may be people's time that you can control by paying them money. And I warm very much to the way in your reply you went straight to evidence. I think people do, perhaps in a new way, want to live evidence-based lives. We talk about evidence-based medicine, and you think, what other type of medicine could there possibly be? It reflects a fresh emphasis on saying we really want to make sure that the interventions, in the healthcare profession, do what they are claimed to do. We really want good evidence that this is a wise thing to do to maximize the probability of helping the person in their given condition; and so I warm to the idea of evidence-based activities and evidence-based interventions, and evidence-based lives. I suppose that then takes you to what kinds of evidence are you willing to take into account, and what kinds of modes of enquiry are you willing to bring to bear on that evidence? So, it may be that in so far as I am saying that this material world is not all, then I am saying that there is lots of other evidence, other kinds of realities that are not amenable to the kind of enquiry that one might otherwise think of as naturalism.

HH: What are you thinking of? What sorts of evidence?

AB: Evidence of friendship. We all experience friendship. Not always, but often, there is evidence. It can be deceptive evidence, it can be just people taking advantage of you, but not usually. A lot of the time it is genuine friendship. Occasionally, the evidence may be material, like a birthday present or something like that, but time and again, it is not something material, and probably the evidence we most highly value isn't material at all in content, though it may be manifested through material means.

HH: OK so let me push on that a little bit harder. Another thing we mentioned in a certain part of the book, in the discussion on quantum mechanics specifically, is this 'intuition to value', a phrase that Andrew Steane used. This is a question for you, then, Andrew: would you count that as a kind of evidence? Do you think, 'I have this feeling there is value' and that is enough for you?

AS: I think that this is evidence; there is such a strong and deep intuition for me, and I think for many others. It seems to me to be the natural way to be human, to suspect that this life is worth something. Those who deny it have, I think, denied it because either there is something that happened to them that I won't go into, or it might be that there's a feeling of caution surrounding what other baggage might be attached to such a statement. There might be baggage which they do not want to pick up, so that they are so cautious about it they end up saying no, there is no value.

HH: I suspect in many cases one of the problems is life experience of a certain kind. People I know who are strongly disinclined to acknowledge the existence of value, value-deniers so to speak, I think often they were taught a value system that they just disagree with and it makes them allergic to ideas that were imposed upon them. As children, they were told here's what's right and here's what's wrong and they found the list of prohibitions to be oppressive.

AS: ... and yet they probably value evidence.

HH: You are right, it's their very sense of value which makes them disavow the values they were taught. Their sense that those aren't good enough, that they have different values from what they were taught so they reject them because they feel that those things are wrong. But the very fact of calling them wrong is a value judgement!

AS: It's like the self-defeating character of logical positivism. 'All processes are meaningless, therefore the things I say or think are not verifiable, including the claims of logical positivism.'

HH: Yes, and I think it is interesting that to be a value-denier strikes me as almost self-undermining.

AS: Oh yes.

HH: If it is so important to you to assert there is no value, why would you even care? Or if you accuse other people of doing something bad by believing in a value, like my value that you should not assert anything without enough evidence, then how have you decided that is bad?

AB: ... like saying, 'I can't stand intolerance!'

HH: I don't want us to lose track of this, since we started out on this with the value of evidence, and I think that is something where, coming out of a background of science, for you that value is important. But how does it actually play out in the practice of science? What are the moral rules of scientific practice? What does good behaviour in a science involve, in terms of evidence?

AS: Well, some of the things are easy to say. When one is tempted just to quietly let drop that little data point that is not quite fitting, when 'we all know that that was the noisy one, just forget it': to start doing that you open up the whole case of falsifying evidence and that's an obvious case. But I think we can come up with more subtle examples.

AB: It differs from field to field. Reporting the data with integrity is part of it. In some sciences, when the rule is you must report all the data, particularly for clinical trials, it is essential that you do. It happens that in some of my scientific work I have done a lot of scanning tunnelling microscopy and frankly most of the data is absolute rubbish and of no value to anybody at all. It's very hard to get the microscope to take good pictures, it's the condition of the tip and the surface and so on. So, if in fact in that field you were obliged to publish all the data, the world would be knee-deep in data that is almost of no value. The skill in scanning tunnelling microscopy is to know when it is as good as it's going to get, and that's when you then record the pictures, because those are the ones you can learn something from. It is very different there from medical trials where data disclosure is essential. Then there is another thing in all the sort of science that I do and that Andrew does. A crucial part of the experiment is that it is repeatable. That was established by Robert Boyle, in seventeenth-century Oxford. He would record in wonderful detail the experiment and which the difficult bits were and how he had to take care so someone else could repeat it and get the same result. In our field, one would say that that is absolutely integral, but it doesn't mean in areas where you can't repeat the experience are therefore not science. If you take the LIGO experiment in 2015, you can't say I won't believe your result until you take another identical pair of black holes and make them collide, please. It's not that kind of science; some of these observational types of science are not amenable to controlled repetition. I suppose that particularly applies to astronomy and palaeontology.

AS: There is a sense in which repeatability plays a great role there in the testing of equipment and bending over backwards to find all ways in which it might mislead you. That involves a massive amount of repeating.

AB: Yes, but that's on the instrumentation, not on the thing you are observing.

AS: I think we already said there are a bunch of other things which are harder to articulate that are to do with evidence and honesty in science and the values of science. It includes giving things a fair hearing. When something is going against the standard story or paradigm you say, 'Well, let's have it, let's hear it', the openness there.

HH: I am trying to think here what other attractions are there. When people use the word naturalism it has many functions that they are trying to indicate about themselves. They are trying to say about who they are (I am a naturalist): how do we advocate naturalism? We picked up some of them. For some people it is literally distancing themselves from certain religious systems or value systems. Then there is also the value of the material world which we agree with and we agree that historically there has maybe been a problem with that: certain people have latched their religious views to material-world-denying. Then there is the emphasis on the importance of evidence in the physical world. A sort of empiricism. Another thing: naturalism often becomes wedded together with a type of modesty. There's more but I'm not willing to say more than what I have in front of me. I think there is an attraction to that but I don't think it is an attraction that is necessarily effective. In fact, a strong version of naturalism actually loses that attraction because it becomes very bald and negating. Things that go beyond what people can see. But I think that a sort of modesty is an attractive attitude to have. Yes, there are some things of which we have a reasonable amount of control, in terms of the gathering evidence aspect, and there are some things when we are more in the dark. I think to be honest about that is a good thing.

AS: I think there was a helpful comparison made by Augustine, and he wasn't the only one, who in the end thought that faith in God is like a romantic thing. Most of the time, we don't go overboard, we share the aspiration of naturalism to modesty, not wanting to make excessive claims. And then what we add to that is a bit like, I hesitate to say it, not to diminish what it is, but there is a sense in which we add a willingness to be a little bit more committed than other people looking on feel is reasonable.

AB: One of the things here that contributes to science is that it is a major part of a scientist's training to be rather rigorous about your degrees of uncertainty. It is something that we all learn, again tracing back to Robert Boyle but maybe it goes earlier still. This is something that exists in other walks of life too. Colleagues who are in business are forever making decisions based on incomplete information and evaluating the uncertainties. They might have a risk register which is a related practice. I think one of the things we have tried to do in the book is to show we are aware of that in the life of faith. We are willing to recognize different degrees of uncertainty in different aspects of our beliefs in a way that we don't find threatening. It may sometimes be a little uncomfortable, but we can live with that discomfort such as it is.

AS: I wanted to add that since we have been considering the role of evidence, a key philosopher was Karl Popper. He showed that one of the thoughts informing scientific enquiry is asking, in any situation, 'What would happen if you were not right?' Or 'What sort of observation would show that you had got it wrong?' That is a question that is very helpful in science. I think it is a question that people who are uneasy about religious faith are wondering. They are really wondering whether we are honest enough to have something that could in principle disprove it, or are we just so attached to it, nothing could possibly happen to dissuade us? There's this worry that we are not honest enough to say, 'If this, thus and that, then we must have made a mistake.' Unless we can say that, what content is there to our saying what we say?

HH: Naturalism faces the same issue of course.

AS: As Christians, we do have some things which we could say. For me it would include such things as, if the celebration of God's love in fact caused people to become less loving, or if it could not embrace cultural diversity, or did not promote fairness and science and high standards in public life, or if it was discovered that the ability to write fiction in a realistic narrative style somehow appeared briefly in first-century Palestine only to be forgotten again, and so on. I dare say we each have our own list.

AB: In Vienna when Karl Popper was growing up there was intellectual activity going on, in Freudian psychoanalysis and in some of the economic theories of the day, that was so constructed that he felt that whatever evidence came along the advocates would embrace that as further confirmation of their theory. That's what made him want to ask this question: what sort of evidence would disprove your theory, and if you can't think of any then it's pseudoscience not real science. But nevertheless, we don't go into the lab in the morning rubbing our hands and thinking, 'What scientific theory can I disprove today?' We are more likely to go in saying, 'I wonder if I can get the experiment to work today; I hope I can.'

HH: I think that is an important point. I was going to draw an analogy with something more general. To start in scientific cases, let's take quantum physics which all three of us have worked with and here's the thing. In a given day you don't want to test, 'Is quantum physics true or false?' but you do want to test to see how it applies to a certain situation. Taking the general theory and filling out a bunch of specific details and seeing if you got that part right. But you know you have a framework and within that framework you have more specific manifestations that you are testing on a daily basis. I think the same is true for the religious person. On a day-to-day basis we are not testing that God exists but we are testing what it means for certain aspects of our lives, and certain theological beliefs are more sensitive to individual situations.

AB: I think that's very helpful and I wonder if we can tie it in to what I was saying just now. We don't go into our life of faith each day saying I wonder if I can disprove that God exists today, any more than we go into the lab wanting to disprove quantum theory. We go into each hoping that we can make it work.

HH: and hoping to understand it better...

AB: ...as a result of trying to make it work and as a result of noticing where it is hard.

HH: Exactly.

AS: I'd like to add, though, there's an added thing here which is liable to be misunderstood but I think it is a genuine aspect of historic Christian faith which is more in the contemplative or mystical side, where every now and again you just say, 'I'm going to let go of whatever is my structure...I'm just allowing myself to acknowledge that maybe everything I've ever thought about God is wrong', and go there just for a while and see what comes back. It's not something to do every day. It's sort of a spiritual exercise. It is not to do with completely uprooting, it is to do with modesty and things like that and yet it gets very close to saying, 'Actually today I'm just going to try and experiment that there is no God'; it gets very close to that. And I think that is a healthy aspect of a Christian life.

AB: Maybe it's a bit like a sabbatical for an academic. For a year, I'm going to try not living in Oxford, I'm going to try living in California. I did that more than once. It's exciting and it's fun and it's stimulating. I don't want to overpress the analogy, but one of things that happened is I then found myself thinking that actually I quite like being back in Oxford.

HH: We have repeatedly said in this discussion that this life is not all. What is that 'not all' which is not captured by the concept of naturalism?

AB: I would be interested to learn different people's answers to that question. Knowing God cannot be contained within naturalism, and maybe knowing other people cannot be either. Both of these are mediated through natural media like showing and speaking and writing, and are experienced through our natural bodies. But if God and other people are beings to be known rather than hypotheses to be tested then that extends beyond what I understand as naturalism.

15

The struggle is nothing new

In this chapter we look at how science and religious commitment work off one another, and we discuss what sort of behaviour is appropriate when there are tensions between them. We will do this by using examples drawn from history. This is often helpful in order to get perspective on debates in one's own era.

The word 'belief' has a slightly different meaning in its standard usage in philosophy from its common meaning in everyday life. In philosophy, 'belief' means purely and simply 'something that is held to be true'. Thus, to say, in a philosophical treatise, that someone 'believes' that rabbits are mammals means simply that that person considers it to be true that rabbits are mammals. The word has no other connotations than that. However, in everyday life the word 'belief' is commonly used to mean a view held without much basis in evidence, or without the ability to present a good case.

In the following discussion we will use the word 'belief' in its technical philosophical sense. We shall mention other meanings in Chapter 17.

When science and religion come into conflict, what is a rational person supposed to do? The default answer among the educated classes in our society is that scientific beliefs are more rational than religious beliefs. But why think that this is so? Some would say that scientific beliefs should be given preferred status because the methods by which they are obtained are superior to the methods by which religious beliefs are obtained. But what reason is there to believe that claim—namely, that science's methods are superior to those of religion? There are two possible sorts of answers. On the one hand, some might claim that the methods of science are *obviously* better than those of religion—that any clear-thinking, rational person can intuit this superiority. We think that this response has little to recommend it. On the other hand, some might claim that the methods of science have proven in the past to be better than those of religion. And what is this evidence that the methods of science are superior to those of religion? It seems that the evidence is supposed to consist largely of cases where religion led to some or other false belief, but that science corrected this belief. So, for example, the ideas of voodoo and animism are called 'religious' and science comes along and displaces them. Or else, one might point to the case of Galileo's scientific argument for the claim that the Earth revolves around the Sun, in contrast to the Catholic dogma of the Middle Ages that the Sun revolves around the Earth. But why call voodoo 'religion'? Why not call it bad science? Or why say that it is science that corrects animism? One could with good insight say that animism is corrected largely by better religion. And is Galileo's experience typical? Does scientific fact supersede and replace religious dogma? And more generally, does science provide an outlook on life that is distinct and incompatible with religious commitment? Or is religious commitment compatible with science in a logical sense, but nevertheless restrictive on the spirit and adventure of science? (You know what we think—it certainly doesn't and isn't! Quite to the contrary, we think that a scientific outlook squares quite nicely with a life of religious commitment.)

In this chapter, we will consider the rationality of religious belief from the point of view of the historical engagement of religion and science. While a commonly touted point of view suggests that there is a battle between religion and science, and science will win,[1] we aver that this story fundamentally misconstrues the growth of scientific and religious knowledge. As scientific knowledge grows, we discover more and more about the physical nature and structure of things. This contributes to our grappling with some of the questions of religion, such as how the world has come about, but it does not tell us what to make of the world, nor how to live, nor how to express our experiences of gratitude and indignance and grief and joy at the way things turn out.

How should we think of the development of science, in particular with reference to its relationship with religious commitment? Many advocates of a scientific outlook believe a narrative according to which scientific knowledge is a replacement for religious belief, and where the scientific outlook is a superior alternative to a religious outlook. As you

---

[1] An extreme statement of this, by the nineteenth-century biologist who introduced the term agnostic, was, 'Extinguished theologians lie about the cradle of every science, as the strangled snakes beside that of Hercules; and history records that whenever science and orthodoxy have been fairly opposed, the latter has been forced to retire from the lists, bleeding and crushed if not annihilated; scotched, if not slain.' Thomas Henry Huxley, *Darwiniana: essays* (1896), 52.

know by now, we reject that view. The scientific outlook is no alternative to a religious outlook, and scientific knowledge doesn't replace the great truths of religion. The point of the remainder of this chapter is to square our point of view with the history of science. To do so, we will point out the problems with the alternative view—i.e. the view that science is triumphantly marching forward to the detriment of religion.

This alternative view is of fairly recent vintage. One of the most vocal proponents of the view was the nineteenth-century social scientist, Auguste Comte. According to Comte, the human race has undergone three major intellectual epochs: the mythical/religious, the philosophical, and the scientific. Comte saw these three epochs as successive improvements, with the philosophical outlook being an advance over the religious outlook, and with the scientific outlook being an advance over the philosophical outlook. For example, Comte saw the Western monotheistic religions (Judaism, Christianity, and Islam) as an advance over polytheistic religions and myths (e.g. Greek and Norse gods). And Comte saw the great philosophical systems (e.g. Aristotle, Kant) as advances over monotheistic religions. And finally, Comte proclaimed that a purely scientific outlook—represented in the terms he himself aspired to show—is the highest form of human intellectual achievement.

Comte's story is neither absurd, nor completely useless. Indeed, we endorse the idea that scientific advances are genuine advances for the human spirit, and we certainly hope that humanity is becoming more rational. What's more, we won't deny that scientific advances are hard-won, and something of which humanity can rightly be proud. And finally, Comte's grand narrative gives an initially plausible framework for some of the major historical interactions between science and religion. However, is it finally plausible? For the purposes of this chapter, let's consider some of the stories often told about how science has gotten the best of religion—told from the point of view in which science supersedes religion.

1. Ancient peoples put forward creation stories ('myths' as we'd now call them) about how the universe was created through copulation of gods, and other such things. But we now know that all of this is nonsense, for the universe has always existed. (This story is told from the point of view of a thirteenth-century scientist who accepts Aristotelian cosmology.)

2. Christians who read the Bible literally are committed to the claim that the Sun revolves around the Earth. But Galileo showed that the opposite is true: the Earth revolves around the Sun.

3. The Western monotheistic religions—Judaism, Christianity, and Islam—claim that miracles have occurred. But we now know that the universe is governed by strict and exceptionless laws of nature such as the Euclidean geometry of space, the conservation of mass, and the continuum of states of energy. (This story is told from the point of view of a nineteenth-century scientist who accepts Newton's laws of physics.)

4. The Western monotheistic religions claim that humans are unique among the animals in that they bear the image of God. What's more, the view that humans are created in the image of God has frequently been accompanied by a claim to the effect that God intervened in the course of nature to create humans—and that humans could not have been a result of ordinary evolutionary processes. However, Darwin's theory of evolution shows that the emergence of the human species was the result of the very same evolutionary processes that produced all other animals.

5. Christianity and Islam claim that there is an afterlife—a claim that is often supplemented by belief in immaterial souls. However, contemporary neuroscience provides an increasingly large amount of evidence that the human mind is inseparable from the body.

Each of these vignettes has been cited—at one time or another—as evidence in favour of the grand narrative that science corrects the false beliefs of religion. And each of these vignettes merits a long critical study in its own right. Here, we will restrict focus to the vignettes drawn from the history of physics. Before we do so, however, let's recall that the historical study of science has grown increasingly sophisticated over the past fifty years. Historians of science would quickly point to shortcomings in this triumphalist account of the history of science. For example, Thomas Kuhn famously argued in *The Structure of Scientific Revolutions* that scientific knowledge does not grow by accretion. In particular, Kuhn showed numerous cases where a firmly established scientific theory was thrown out in favour of a new theory. Kuhn thought that the transition from old to new theory was so abrupt that he called it a 'scientific revolution'—with the specific implication that the old theory and its knowledge claims are deposed.

So, when a religious person is faced with a potential conflict between her background beliefs and the claims of science, should she take solace in a Kuhnian view of the history of science? Should she think, 'No need to worry about the conflict, because the current scientific paradigm will eventually be superseded'?

We think not. For one, Kuhn's radically anti-cumulative view of scientific knowledge is not the only feasible way to understand what happens when one scientific theory replaces another. In particular, a new theory could be a direct improvement upon an older theory in the sense of being 'closer to the truth'. At least that is the idea put forward in the article 'The Relativity of Wrong' by Isaac Asimov.[2] According to Asimov, the growth of Western science is a story of ever closer approximations to the truth. For example, science taught us that the Earth is a sphere—which is not strictly speaking true, since not every point on the earth is equidistant from its centre. But it's more accurate to say that the Earth is a sphere than it is to say that it's flat.

Asimov's article is merely a catchy example of a widely held point of view of what science is supposed to do for us. It by no means provides the view's clearest articulation, or its strongest defence. Let's call this view *convergent realism*, borrowing a term from philosophy of science.[3] According to convergent realism, science progresses by moving our beliefs into closer alignment with reality. For example, 'The Earth is flat' is replaced by 'The Earth is round.' Particularly relevant for us is the way that convergent realism is often pressed into service in discussions of the relationship between science and religious belief. In particular, there seems to be a widespread misconception that religious beliefs can be nothing more than primitive attempts at answering scientific questions.

It's an appealing picture, where human questions remain the same throughout the ages, and science provides the most effective means of answering these questions, or at least of moving us closer towards the answers. But this idea just doesn't square with our experience of science, or with what we know about its history. In fact, science frequently tells us: 'Your question is too fraught with meaning for me to answer. If you want my help, you will need to pose a different kind of question.' In

---

[2] Isaac Asimov, 'The Relativity of Wrong', *The Skeptical Inquirer* **14**:1 (1989), 35–44.

[3] See L. Laudan, 'A Confutation of Convergent Realism', *Philosophy of Science* **48** (1981), 19–49. Our concept of 'convergent realism' has affinities to the concept discussed in Laudan's article and elsewhere in the philosophical literature, but is not intended to be completely faithful to that usage.

the remainder of this chapter, we will discuss two historical cases where science *refused* to answer the original question. In other words: science did not give a new answer to the same old question. First, we will look at a case from the thirteenth century, when people were asking science: Did God create the universe? Then we will look at a case from the seventeenth century, when people were asking science: Does the Earth move?

15.1 Thirteenth century

Imagine living in Europe in the thirteenth century. Most likely you would have been Jewish, Christian, or Muslim. In any case, if you were educated—which in and of itself would have been unlikely—then you would probably have been a member of a religious order. And your academic training would have consisted primarily of studying Scriptures, and also the writings of the 'fathers' of your religious tradition. You would have learned how to cite these authorities in defence of whatever claims and counter-claims you considered in a discussion of points of law or philosophy, a practice known as scholasticism.

One of the claims in Scripture is that God created the heavens and the Earth. The fathers of the Western monotheistic traditions took these statements to mean that the universe was created *in time*, i.e. that it was *not* eternal. More precisely, they upheld the doctrine of *creation ex nihilo*—that God created the world out of nothing. One might say that time itself, or temporal existence, was created, along with everything else that was created, and by now some finite, not infinite, amount of time has passed.

The reason for belief in creation ex nihilo was probably some combination of: (1) taking Scripture as an accurate description of historical facts, (2) trusting the interpretations of religious authorities, and (3) philosophical reflection (in particular, thinking about what it means to say that 'God created all things'). Empirical evidence plays some role in supporting this belief. People routinely observe that material things (e.g. flora, fauna) have finite lifespans and depend on other things for their existence. Thus, belief in creation ex nihilo has probably also, in some cases, been motivated by a wrong-headed attempt to use God as an explanatory posit on the same level as other scientific explanations.

We said that an educated person in thirteenth-century Europe would have relied upon the testimony of religious leaders, in particular certain notable figures in the history of his tradition. However, in the twelfth century, Christian scholars came to possess the works of the great Greek philosopher Aristotle. To call Aristotle a 'philosopher' is a bit of an anachronism, for Aristotle was a polymath. Indeed, Aristotle can with rights also be considered the first scientist in the Western tradition, for one of his primary goals was to describe and explain the full range of empirical phenomena.

The religious philosophers of the twelfth and thirteenth centuries were both impressed and vexed by Aristotle. They were impressed because Aristotle could offer explanations for many things—why rocks fall, why trees grow, and why humans seek friendship. But they were vexed, because if you accept everything that Aristotle said, then you will also end up rejecting some parts of religious doctrine. There were three main areas of conflict between religious doctrine and the Aristotelian worldview.

First, Aristotle famously rejected Plato's dualistic picture in which the material world is a shadow of a more perfect, transcendent, non-material world. By the twelfth century, Plato's view had thoroughly penetrated Christian thought, so that the Christian doctrine of the afterlife came to be regarded in the light of Plato's doctrine of immaterial (non-perishable) souls. But then what is one to do when Aristotle says that Plato was wrong—that the human soul is not an immaterial object, but is rather the form or function of a material being? To make things worse, Aristotle's account of matter and form strongly suggests that all human beings share the same soul, leaving no room for individual existence in an afterlife. Thus, there was an apparent conflict between Aristotle's theory of the soul and the Christian doctrine of the afterlife.

The second major area of conflict between Aristotelian thought and Christianity was in the realm of ethics. Christians claim that the highest good for mankind is to be in a right relationship with God. And this traditional doctrine seems to justify a couple of other claims about how to live the best kind of life. First, it was thought that the fullest communion with God cannot occur until the afterlife—and hence that the greatest good for humans cannot be realized in this life. Second, it was thought that the best way for a human being to spend their time in this life was in worshiping God and serving Him.[4] In contrast to these

---

[4] This combination is capable of great creativity and human sympathy—it should not be taken to mean merely religiosity.

commonly held beliefs among Christian thinkers, Aristotle claimed that humans are intrinsically or primarily rational beings, and so their greatest good consists in their exercising their rational faculties. For Aristotle, the highest form of existence is one that consists in pure thought. By contrast, the Christian community inherited a combination of both Greek and Hebrew ideas, and the latter saw the relationship to God as more like a partnership or covenant, in which God is concerned for the plight of everyday people and the injustice they face, and we are called to notice and respond in practical ways.

These first two conflicts between Christian teaching and Aristotelian philosophy deserve extended study in their own right. However, we will be concerned with a third conflict, which involves an incompatibility between Christian doctrine and Aristotelian physics.

In essence, the conflict is that Aristotelian physics says that the universe had no beginning, i.e. that it has always existed, which is in direct contrast to the doctrine of creation ex nihilo. Thus, if we were to over-simplify things, we might simply describe the conflict as follows: religion said that God created the universe such that it had a beginning, and science said that the universe had no beginning. Of course, then, if you were living in the thirteenth century, and if you applied the maxim 'always trust science over religion', then you would have believed that the universe had no beginning.

Before drawing any implications from this case, let's remember more particularly why Aristotle's physics was committed to the claim that the universe is infinitely old. For this claim was no accidental feature of his approach. It could not have been abandoned without massively altering the very foundations of physics—something that the medieval thinkers were hardly in a position to do.

Some people would say that Aristotle laid the conceptual groundwork for what we now think of as science, and that before Aristotle's conceptual breakthrough science as we know it was inconceivable. Evaluating that claim is beyond the scope of our book. However, we can at least point to a few of the reasons why one might think it's true. Perhaps the main reason is that before Aristotle, Greek thinkers were caught up in the dichotomy of spirit and matter, with many of them—such as Plato—attributing a higher level of reality to the immaterial word, what we now call 'Platonic forms'. The world of matter was seen as constantly changing, which meant that it wasn't really possible to have knowledge of the world of matter, since knowledge requires a sort of stability that cannot be found in the material world. Thus, while Plato's theory of immaterial forms provided a nice explanation of mathematical and even ethical knowledge, it seemed to undercut any motivation for empirical investigation.

In contrast to Plato's picture, Aristotle's picture invested greater significance in the material world. In fact, it might be said that Aristotle was a sort of materialist, since he didn't believe that there was an immaterial world that was equally or more real than the world we see around us. And yet, Aristotle maintained that it was possible to know things about the material world. How could this be so if the material world is constantly changing? What Aristotle said is that in every change, something stays the same. And this thing that stays the same, we can call *matter*. Consider an example: a person grows an inch taller over the period of one year. We can analyse this example into two components. First, there is the underlying person, who is the same from one year to the next. Second, there is the height of the person, which increases from one year to the next. We can think of the underlying person as the *matter*, and we can think of the person's height as the *form*. Thus, for Aristotle, every change involves a change of form in the same underlying matter.

The notion of 'change' lies at the very foundation of Aristotle's physics. He explains every event as the result of certain changes—in particular, changes in which an object fulfils its natural role, and changes in which an object is prevented from fulfilling its natural role.

This being so, it's not difficult to see that Aristotle's physics is incompatible with the universe coming into existence a finite amount of time in the past. The style of argument we'll use here is *reductio ad absurdum*, where supposing the opposite is shown to lead to an absurdity. So, suppose that the universe came into existence at some moment t_0 which occurred at a finite amount of time before the present. Then matter existed after t_0, but did not exist before t_0. But the change (from nonexistence to existence) requires an underlying substratum, and so something—some material element of the universe—existed before time t_0. This is a contradiction, which shows that the supposition must have been false, and the universe had no beginning.

The thirteenth-century philosophers faced the following dilemma: on the one hand, the Scriptures teach that the universe was created by God and is not itself eternal. On the other hand, reason, i.e. science,[5]

---

[5] We will refer to Aristotle's body of work by using simply the word 'science'. This is because it played a role, for thinkers at the time, very similar to the role that science plays for us now.

teaches that the universe has always existed. So, if these philosophers were being rational, then wouldn't they have abandoned their religious belief that God created the universe?

The conflict between the doctrine of creation ex nihilo and Aristotelian cosmology occupied the minds of the sharpest thinkers of the twelfth and thirteenth centuries. And these thinkers by no means agreed on what ought to be done about this conflict. In fact, the responses to Aristotelian physics took three basic forms: (1) reject Aristotelian physics because it conflicts with revealed truth; (2) reconsider the foundations of Aristotle's physics, and try to show its consistency with revealed truth; and (3) accept Aristotle's physics and argue that there is no genuine conflict with revealed truth. Let's go through these three sorts of response in turn.

The first sort of response found its most clear voice in a Franciscan monk, now known as St Bonaventure. As with the other characters at this point in our story, Bonaventure was a master at the University of Paris—which meant that he was a sort of hybrid theologian, philosopher, and proto-scientist. But Bonaventure was more concerned than most about maintaining doctrinal purity; and he was more of a Platonist mystic than an Aristotelian scientist. In any case, Bonaventure thought that if Aristotle's physics entailed that the universe was eternal then there was something wrong with Aristotle. What's more, Bonaventure claimed that what was wrong with Aristotle is that his 'Reason' had been darkened by sin, and needed to be illuminated by God's light. Finally, Bonaventure claimed that 'true Science' showed that the universe had not always existed, and in fact that it had a Creator.

Before proceeding to the next figure in our story, it might be illuminating to compare Bonaventure to contemporary religious thinkers who find their beliefs in conflict with the prevailing scientific paradigm. Think, for example, of the so-called 'creation scientists' of the 1970s and 1980s, or of the intelligent design theorists of the twenty-first century. In both cases, the claim is that the prevailing scientific paradigm has gone astray because of some implicit bias—in particular, a bias toward naturalism. They claim further that when this bias is eliminated, then better scientific theories can be developed. And, oh, coincidentally, these better scientific theories point to a creator.

Bonaventure's proposal was thoroughly rejected by the most powerful mind of the Middle Ages: St Thomas Aquinas. In his famous discourse *The Eternity of the World*, Aquinas points out numerous fallacies in Bonaventure's supposed proof that the universe has a beginning.

Where did this leave Aquinas? If Bonaventure's arguments are faulty, then is it irrational for the religious believer to maintain that the universe was created by God? No it's not, said Aquinas. Rather, Aquinas argued that not only is Bonaventure's argument for a finite universe flawed, but so is the Aristotelian argument for an eternal universe. Indeed, Aquinas deemed that reason (or, what's the same thing in this context, science) was simply incapable of determining an answer to the question: Has the universe always existed?

This should not be interpreted as Aquinas stating a view on whether the physical universe, or some sort of physical substratum, has always existed; he is stating simply that the application of scientific reasoning is incapable of deciding that question. In a modern expression, his argument would say that, notwithstanding the fact that there is plenty of evidence that time as we know it started off in the cosmological Big Bang, this does not enable us to know, from science alone, whether there was a physical thing which predated the Big Bang and gave rise to it, and which could be said to always exist or always have existed. That sort of question is in a different category from those addressed by reason (i.e., roughly speaking, the scientific method), says Aquinas.

This bold move by Aquinas was not appreciated by all religious thinkers of his time. And yet, Aquinas' thought on this issue has proven to be powerfully influential—and dare we say, even fruitful for the development of science. For Aquinas thought there were issues of metaphysical importance—such as whether the universe was created—that simply transcend the power of science. And that meant that science could be allowed to develop, without fear of negative repercussions for these transcendent issues.

There's much to be discussed about Aquinas' response to the conflict between creation ex nihilo and Aristotelian physics. However, let's simply note for the time being the similarities and difference between Aquinas' view and that of the contemporary evolutionary biologist Stephen Jay Gould, who (as we saw in Chapter 3, and shall mention again later) claimed that the domain of science is fact, and the domain of religion is value. For Aquinas, the domain of religion—or faith—does include factual claims, most relevantly, the claim that God created the universe. There may be some way of reconciling Aquinas' view with that of Gould, but for now we move on to our third group of thinkers trying to reconcile Aristotelian physics with religious faith.

Back in the twelfth century, Islamic scholars had already been struggling with integrating the insights of Aristotelian science with their religious doctrines. Among the various responses that were developed, the most radical was that of Ibn Rushd, also known as Averroes. At the opposite end of the spectrum from Bonaventure, Averroes believed that Aristotle was essentially right—about almost everything. So when Aristotle 'proves' that the universe had no beginning, Averroes says: 'So it must be.' But what then happens to the doctrine that God created the universe? Here Averroes says something quite mysterious: he says that *qua* philosopher, he believes that the universe is eternal, but that *qua* Muslim, he believes that the universe had a beginning. This way of talking about the issue came to be known as the *doctrine of double truth*.

Averroes had a significant impact on Western thought—and, in particular, on the secularization of Western thought that occurred in the late Middle Ages. In the medieval Christian world, Averroes was widely regarded as an acute commentator on Aristotle's work, although many Christian philosophers regarded him as a radical, essentially denying many of the core doctrines of the faith. There were, however, some Christian Averroists—the most notable being Siger of Brabant and Boetius of Dacia, who were also teachers at the University of Paris, along with Bonaventure and Aquinas.

According to Siger and Boetius, doing science involves adopting a set of rules—e.g. rules about what a good explanation will look like. Siger and Boetius claimed that the locution '*qua* philosopher, I believe that X' can be cashed out as saying, 'According to the rules of science, with no recourse to further information, X is the most reasonable conclusion.' In other words, to say that, '*qua* philosopher, I believe that X' means that 'X is true within the game of science,' and not necessarily true in some absolute sense.

There's no doubt that Siger and Boetius were espousing a form of relativism about truth. And many have dismissed their views—even called them heretical—on this account. The views of Siger and Boetius were condemned by the Catholic Church in 1270.

Let's try to be fair to Siger and Boetius, and let's remember that not all relativism about truth is pernicious. Consider, for example, the claim that:

The sum of the internal angles of a triangle is 180°.

Is this claim (absolutely) true? It would be difficult to maintain that it is, in the light of the discovery of non-Euclidean geometry in the nineteenth century. For, in systems of non-Euclidean geometry, it is *not* the case that the internal angles of a triangle sum to 180°. But is there reason to say that Euclidean geometry is itself absolutely true? This question is difficult and has exercised the minds of great scientists and philosophers such as Henri Poincaré, Albert Einstein, and Bertrand Russell. According to these thinkers, the best we can do is say that the claim (that the angles sum to 180°) is true relative to Euclidean geometry, and false relative to non-Euclidean geometries. Indeed, Poincaré went so far as to say that geometrical claims are *conventional*, and hence cannot be (absolutely) true or false.

Now consider again the claim that the universe is eternal. Could it have been rational for Siger and Boetius to say that this claim is true *relative to* Aristotelian physics, but false *relative* to Christian faith? Isn't there a strong disanalogy between the claim about the age of the universe and the claim about the internal angles of a triangle? After all, the triangles of mathematical geometry are ideas in our head, and we are free to define them as we wish (so long as our definition is consistent). On the other hand, when we're talking about the universe, we're presumably talking about something outside of us, whose features don't depend on our making any definitions. So isn't it a purely factual question whether the universe is eternal?

The disanalogy here is deceptive. On the one hand, if you draw a triangle on a piece of paper, then it's just as much a part of physical reality as the universe—and so the question of whether its internal angles sum to 180° is just as much a factual question as the question about the age of the universe. On the other hand, the notion of 'the universe' in physical science is itself a theoretical posit, and applying the notion of 'infinite age' to this theoretical posit presumes some method of correlating our numerical concepts with events in the world. In either case, the disanalogy between the claim that a triangle's angles sum to 180° and the claim that the universe is eternal is not as serious as one might think. In both cases, the claims occur within a system which presupposes certain rules and definitions—and so in both cases, it's possible to talk about relative truth without falling into incoherence.

Thus, when Siger and Boetius say that 'the universe is eternal' is true *relative* to Aristotelian physics, they were way ahead of their time in understanding the nature of physical theory—for they say that a physical theory is a rule-governed system, not merely a collection of unrelated facts. In particular, one cannot both accept Aristotelian physics—

which is based on the idea that all change requires a substratum—and say that the universe had a beginning. For the very notion of 'beginning', or a time before which there was no substratum, is inconsistent with the foundational concepts of Aristotle's system.

But shouldn't a Christian with integrity then reject Aristotelian physics? It's tempting to say yes. But Christians also have a high stake in promoting the development of scientific knowledge. And here Siger and Boetius have another insight: scientific paradigms are unified entities, and sometimes we must accept some unpleasant features (e.g. conflicts with our other beliefs) in order to obtain the numerous other benefits (the systematicity, predictive power, etc.). And it seems that Siger and Boetius were aware—in a way that Bonaventure was not—of the virtues of the Aristotelian paradigm, and of how difficult it would be to find an equally powerful paradigm without the consequence that the universe is eternal. So what's a Christian thinker supposed to do? It seems in this case that Siger and Boetius did an honest thing: they admitted that they wanted the power of Aristotelian physics, and they wanted to maintain Christian faith, and they simply couldn't see how to make it all fit together in a neat package.

In summary, then, the medieval Christian philosophers staked out three attitudes toward a physical theory (Aristotle's) that conflicted with their religious doctrine (creation ex nihilo):

1. Propose a 'better' scientific theory that is consistent with religious doctrines (Bonaventure—compare with modern-day intelligent design theorists). The danger here is that the alternative theory might be worse, in many ways, than the original theory. A scientific theory can be 'good' along many different dimensions—being simple, powerful, beautiful, systematic, well integrated with other theories, etc.
2. Argue that the science doesn't reach as far as some scientists say it does, and in particular, it doesn't impinge directly on theological doctrines (Aquinas, and compare S. J. Gould).
3. Argue that science speaks a somewhat different language than theology, and hence apparent conflicts can be temporarily embraced (Siger and Boetius, and compare S. J. Gould).

Which of these responses do we think was the most reasonable? While we are inclined to censure Bonaventure—both for his sloppy thinking and for his fear of science—we understand his impulse to defend the truths of religion. The problem with Bonaventure's approach is that he seems to invest too much faith in the power of science to settle theological questions. Unlike Aquinas, Bonaventure thinks that science has something to say about whether the universe was created. We are also inclined to praise Aquinas, Siger, and Boetius for their faith that an apparent conflict between science and religion would eventually be ironed out. Such faith is not unreasonable, and indeed was a trust held by them with care and attention to their reasons for it and how it should be handled. And history has vindicated their faith—for contemporary science sees little to no validity in Aristotle's argument for an eternal universe.

We're also inclined to praise Aquinas, Siger, and Boetius for taking a more laissez-faire approach to scientific theories. In essence, they said,

> Science is a serious game, and a scientist should be allowed to play this game on its own terms. Science isn't trying to answer the same questions as religion; and if it seems to provide a conflicting answer, then we may, with integrity, choose to trust that someday the apparent conflict will be shown to be spurious.[6]

Put more simply, the eternity of the universe episode shows that purported conflicts between science and religion can have more to do with our limited insight than with inconsistency. In this case, the medieval thinkers could do no better than to identify reason with Aristotelian thought—it was the best model that they had. And by the lights of Aristotelian rationality, it wasn't rational to conclude that the universe had a beginning. But Siger and Boetius were smart enough to see that Aristotelian science provided only relative truths, albeit truths that they ought to recognize since they had no superior system to which they could turn. Perhaps this case from medieval cosmology can provide some insight to those who see conflicts between contemporary science and even the best-motivated religious belief. The scientific truths of today are also time-bound. And if science grows in the way it ought to grow, then our descendants in six or seven hundred years will see that our perspectives were limited—and they may judge that what we saw as a conflict was, in fact, due to limitations of our perspective.

Again, we want to be clear that we are *not* suggesting that science and religion cannot conflict; or that, if they do conflict, then we ought to

---

[6] The words here are our own, but our terminology borrows from the acute analysis of Sten Ebbesen, 'Boethius of Dacia: Science is a Serious Game', *Theoria* **66** (2000): 145–58.

take a blasé attitude to the effect that 'it will all work out in the end.' It is our responsibility to manage our system of beliefs. And inconsistency is most certainly a severe defect in a system of belief. Perhaps what we're trying to say here is that inconsistency—or, at least apparent inconsistency—is not necessarily a fatal flaw. Sometimes we might have very good reasons to hold onto two beliefs which appear to be inconsistent with each other. In such a case, we *should* feel uncomfortable, and we *should* attempt to restore consistency. But we should not make the same mistake as Bonaventure in denying that there is a good reason to hold onto the deliverances of our best scientific theories.

15.2 Seventeenth century

Let's now turn to a better known episode in the history of science—namely, Galileo's argument for a heliocentric universe. This episode is typically taken as the paradigm case where religious dogma stood in the way of the growth of scientific knowledge. In particular, it's often thought that Galileo tried to replace a false belief (the Sun moves around the Earth) with a more true belief (the Earth moves around the Sun). But the actual facts of the Galileo case are more subtle than that. In fact, the moral of the Galileo episode is that science sometimes refuses to answer the questions that we originally pose to it.

In the following we are not going to present a historical analysis of the legal and moral aspects of the dispute between Galileo and officials of the Roman Church in the seventeenth century. It is appropriate to admit that Galileo was badly treated by some in the church hierarchy, and supported by others in that same hierarchy. It is appropriate to admit that the Roman Church developed laws around censorship that were too heavy-handed, while nevertheless continuing its active support for astronomy through the work of Jesuits such as Giovanni Battista Riccioli. Overall, one could make a case that, notwithstanding his faults, Galileo here better represents what it means to be a follower of Jesus (for that is what he was) than did various clergymen.

But the moral and legal aspects have been discussed elsewhere. Here we want to draw attention to an aspect that concerns purely the history of ideas, and the impact of science and religion on how we think and understand.

The question under consideration in seventeenth-century astronomy was whether the Earth is in continual motion as a whole, and what was the right overall model for the locations and motions of astronomical bodies such as planets, Sun, Moon, and stars. Related to this was the question "How can we best find out about such things, and get reliable knowledge, without inviting the kind of social upheaval that results in large amounts of destructive violence?" (The latter consideration was very much a live issue at the time.) What we want to do here is show that the question about the motion of the Earth was not resolved simply by replacing an untruth with a truth. The statement 'the Earth does not move' looks like a direct contradiction of the statement 'the Earth does move', but from the point of view of modern physics, the relation between these statements is not quite so mutually exclusive or contradictory as it appears. When we point this out, we are not seeking to either exonerate or condemn any decision taken in the past; our aim is merely to show that sometimes what looks like a direct confrontation between scientific ideas and religious concerns can turn out to be resolved not by one side having to abandon what they value, but by both sides arriving at a coherent larger view, in which both have learned something valuable.

To see what's meant by these claims, recall that according to modern physics there is no fact of the matter about whether you are really moving or are stationary. In other words, motion is relative. Consider an example: suppose that you are taking a New Jersey transit train into New York Penn Station, and the train is in the dark tunnel that leads into the station. As anyone who has ridden this train knows, there are frequent delays, and often the train will sit for several minutes in the dark tunnel leading into the station. Travellers on the London Underground can have similar experiences. The tunnel is so dark that it's not possible to see the tracks, nor the side walls. And so at times, it's simply impossible to tell whether your train is moving. Sometimes, however, if you look out the window, you will see a train moving in the opposite direction. But is the other train standing still and is your train moving? Or is your train standing still and the other train moving? Or are both trains moving?

The example we've just given is analogous to a famous thought experiment by Galileo. Galileo asks us to consider the following scenario: suppose that you're on a ship that is sailing at a constant velocity in one direction. Suppose, however, that you cannot see the shore. In fact, for the purposes of the example, we may suppose that you are in a closed cabin with no windows. Now, is there any experiment that you could do in order to detect whether you are moving (relative to fixed land)?

The answer, says Galileo, is no—the laws of physics are the same in all inertial reference frames.

Galileo's insight about the equivalence of all inertial reference frames wasn't fully implemented in physics until Einstein's special theory of relativity, though Einstein had to abandon what had become known as the Galilean transformation. Nonetheless, the insight was already there—and it has proved to be one of the most profound insights in physics.

If you've followed us so far, you'll agree that inertial motion is undetectable. But what about accelerated motion? That is, is it possible to detect whether you are accelerating? Here the empirical data seem to speak in favour of the detectability of accelerated motion. Think again of the example of the train rolling into Penn Station. Suppose that you have a cup of hot coffee in your hands, and you're going to use your cup of coffee as a motion detector. As we argued earlier—in agreement with Galileo's ship example—your coffee can't tell you whether your train is moving, i.e. whether it has nonzero velocity relative to the ground below your feet. However, every coffee drinker knows that acceleration and deceleration are dangerous: if the train suddenly lurches forward, then your coffee is likely to slosh out of the cup. Similarly, if the train comes to a sudden stop, then your coffee is likely to find its way onto the floor or your shirt. Thus, the data of everyday life seem to speak in favour of the reality of absolute acceleration.

Let's also recall that circular motion is accelerated motion. Think, for example, of a car driving around a circular track, and imagine an open cup of coffee in a passenger's hands. Again, every coffee drinker knows that in this situation, there is a danger that the coffee will slosh out of the cup sideways and onto the floor of the car. In short, the coffee in the cup would naturally continue motion on a straight line, so when the car bears to the left or the right, you can detect this motion by looking at your cup of coffee.

These sort of thought experiments suggest the idea that there is a fact of the matter about which object is orbiting which when objects are in orbital motion. In reference to our main example: there is a fact of the matter about whether the Sun is orbiting the Earth (geocentrism), or whether the Earth is rotating about its own axis and orbiting the Sun (heliocentrism). And according to the 'triumphal march of science' story, religion stated falsely that the Earth is the centre of rotation, whereas science states truly that the Sun is the centre of rotation.

This story might seem convincing if you know a little bit of physics. However, when you learn more physics, you see that it's at best a half-truth. For Einstein showed that acceleration is also relative, and hence that orbital rotation is relative.[7] In other words, there is no fact of the matter about whether the Earth or the Sun is the centre of our Solar System. To be fair, we must admit immediately that there is no doubt as to which picture is the clearer and more natural for the purpose of analysing planetary motions, but we are here making a technical point about accelerated motion in order to build towards our conclusion, which will be that the situation is more subtle than either party could have known in the seventeenth century.

How could Einstein be right in saying that acceleration is relative when our coffee tells us otherwise? Here was Einstein's great insight—the so-called principle of equivalence: it's impossible to detect whether you are accelerating, or whether you are avoiding acceleration in the presence of a massive object. And here's Einstein's thought experiment meant to convince us of the principle of equivalence: imagine yourself in a rocket ship, again with no windows. Now suppose that what you feel is a force that pulls you down toward the floor—the same sort of force you feel when an elevator accelerates in upward motion. There are two hypotheses consistent with this data: one hypothesis is that the rocket ship is in empty space and is accelerating in the direction that you would call 'up'. The second hypothesis is that the rocket ship is sitting on the launching pad and isn't moving at all (relative to the Earth). In other words, the force of gravity pulling you down to the floor is completely indistinguishable from the effects of acceleration. This thought experiment, Einstein claims, indicates that accelerated motion itself is relative—there is no absolute fact of the matter about whether you are accelerating. You can only determine whether one body is accelerating relative to another.

This conclusion is so unfamiliar that it can be hard to accept it, so let's clarify. Einstein's insight invites us to a statement about acceleration

[7] In the context of the general theory of relativity, the question of absolute rotation can be posed as follows: in terms of the objective features of the theory (e.g. space–time metric, stress-energy tensor), can one distinguish rotating objects from non-rotating objects? The answer to that question is no. See Section 3.3 of David Malament, *Topics in the Foundations of General Relativity and Newtonian Gravitation Theory* (University of Chicago Press, 2012). We humans are free to designate some objects as stationary, and others as rotating, but such designations are not a part of the human-independent structure of physical reality.

similar to the more familiar statements about velocity. Two parties can say whether there is mutual acceleration between them, but they cannot claim to know absolutely which of them is accelerating. When one party detects evidence of what they interpret as acceleration, such as spilling coffee, it might equally be a case of gravitational force. That is, the coffee in your cup might have spilled not because the train (and your arm) accelerated, but because it refused to accelerate when a local gravitational field invited it to do so. Having said that, we admit, and Einstein would also admit, that this does not change the fact that in many everyday situations one perspective can be more natural and useful than another.

Finally let's return to our main question, whether Galileo replaced a religious dogma (geocentrism) with a scientific fact (heliocentrism). One problem with this claim is that it is unhistorical. The religious dogma of the day was that heliocentrism could be discussed and used as a calculational method, or an appearance of things, but should not be presented as a fact of the matter, because the evidence did not warrant that. Galileo asserted that the Earth moves. The latter view became established when more evidence was accumulated—scientific work that was done as much within the Church as without it, with notable contributions from Giovanni Battista Riccioli in the seventeenth century, and the observations of The Revd Dr James Bradley in 1728.[8] But what eventually became clear is that in the original dispute both parties were partly right, both partly wrong, about the physical science. The relativity of motion extends even further than Galileo suspected, and modern physics has profoundly shifted the perspective. In a technical but completely well-defined and accurate sense, it is no more true to say 'the Earth moves' than to say 'the Earth does not move'.

From the point of view of modern physics, with general relativity 'in our back pocket', as it were, the dispute that occurred in the seventeenth century looks very different. We bring in the idea of a frame of reference, and we begin by noting two things. First, we can formulate physics so that the same equations apply irrespective of the frame of reference, and this makes the choice of frame of reference largely arbitrary. Then we also note that, notwithstanding this, the cosmos offers, at the largest scale, a choice of frame which is a natural choice for looking at the big picture of the universe: it is the reference frame in which the distant galaxies are not moving on average (this is called adopting 'co-moving coordinates' in discussions of physical cosmology). It is this reference frame which we typically pick in our mind's eye when we sit above the plane of the Milky Way galaxy, as it were, and look down at the Solar System from far off. In this frame the Earth is certainly rotating about its own axis, and the distant galaxies are not orbiting planet Earth, and neither is the Sun for that matter. However, this perspective does not make us at all troubled by statements such as 'the Sun rises', 'the Sun moves', and so on. When our fellows say, of their experience, 'look, the Sun rises, moves across the sky, and goes down', then their statement is perfectly fair and correct, and physics has not proved them wrong. It is a correct statement of what goes on in another perfectly natural choice of reference frame: the one that is natural to describing everyday life on any region of the surface of the Earth.

Although the trial of Galileo involved some valid disputes about the nature of Scripture and the care needed not to overstate evidence, in one telling respect Galileo and the Church authorities were fighting over nothing. Their dispute was no more meaningful than a dispute about whether weight should be measured in pounds or kilograms. In both cases, what physics tells us is that there is no fact on which this dispute could turn.

Don't get us wrong. We are *not* claiming that other Church figures were right to attack their fellow Christian Galileo, and we are *not* claiming that it would have been rational for people to continue believing that the Sun revolves around the Earth. Not at all. Rather, we are saying that the Galileo episode is frequently oversimplified as a case where a false claim was replaced by a true one, when, in fact, all parties were equally feeling their way about what science could say about the motions of heavenly bodies. What we've learned since then is that physics has no stake in a heliocentric versus geocentric debate; it rejects the terms of that debate.

Now the reader may want to say that what we are arguing here is missing the point. What is important in the Galileo case is that he was arguing for the value of free enquiry, while Church authorities were trying to control or constrain free enquiry. No doubt, there is some truth in this statement. But our aim here is not a full analysis of the history, but merely to bring out one aspect.

Advocates of the 'triumphal march of science' narrative would have us think of Galileo speaking as follows: 'The Earth moves. For the good of humankind, the truth of science must prevail over the myths of religion.' But this construal would make Galileo look silly, for then he would have fought simply to replace one belief, based on a false presupposition, with another belief, based on the very same false presupposition—namely, the presupposition that there are absolute (i.e. not relative to frame of reference) facts about which bodies are moving.

Perhaps, then, we can construe Galileo as saying, 'we are allowed to consider and teach that the Earth moves.' This is alright, and it is what everyone has come to agree. But it involves a widening of the notion of the word 'moves' beyond its meaning in everyday speech, and in this widening something different is happening than the mere suppression of religion. In everyday speech it is quite correct to say that the Earth does not move, after all. When we say 'it doesn't move' we mean it is still here, it does not suddenly whizz off somewhere else, or crumble or wobble when we jump up and down.

Perhaps we can think of the Galileo episode along the lines which suggest that while science typically produces claims that will be seen to be false (from the point of view of future science), still, the claims it produces are getting closer and closer to the truth. So, is it correct to think that heliocentrism is closer to the truth than geocentrism?

No: this is *not* the take-home point of the Galileo episode—that heliocentrism is more true than geocentrism. To reiterate the point we made before, if we take heliocentrism to mean that 'the Sun is stationary and the Earth rotates around it', then heliocentrism is not even false, but is meaningless—because we now know that phrases like 'X is rotating around Y' have no absolute meaning. They only have meaning within a coordinate system, and no coordinate system is uniquely privileged to describe all phenomena. Rather, any coordinate system can be used, and in practice one system or another is more useful for one purpose or another. The coordinate system most natural to describe large-scale physical cosmology is the one in which large scale structure in the universe is not moving. The coordinate system most natural to describe everyday experiences and experiments on Earth is one in which the immediate surroundings are not moving.

Galileo's key scientific contribution—if we may interpret him charitably, from the point of view of current science—was to say that we need not privilege the frame of reference in which the Earth is stationary.

It is equally valid to work in a frame of reference in which the Sun is stationary (and, in fact, the relative motions of bodies in our Solar System can be described by means of simple and elegant equations in this frame of reference). Seen from this point of view, it's clear why Galileo believed that his claims about the Solar System were consistent with accepting the biblical stories as literally true. The Bible reports accurately the experience of living on planet Earth and observing the motion of the Sun. With the development of science we learn to call this relative motion. The Scripture remains accurate in its reporting of the form this motion takes when observed from Earth.

Galileo thus took a stance similar in many ways to the stance taken by Aquinas, Siger, and Boetius several centuries earlier. Indeed, Galileo himself never saw his discoveries as undermining the Catholic faith itself—he only saw them as undermining a mistaken view about how religious doctrines ought to be integrated with empirical science. For Galileo, as for Aquinas and Siger, empirical science has a particular kind of aspiration. It's not aiming to find truth *tout court*; rather, it's aiming to find those truths that are amenable to systematization and mathematization. And we don't find in Galileo any claim to the effect that the truths of empirical science are an adequate replacement for the truths of religion—truths such as that we receive life as a gift, and we serve but do not own the ecology of the Earth, and we find the fullest expression of our humanity in humble communion with God. Quite to the contrary, it seems that Galileo's numerous scientific successes were bought at the price of aspiring to achieve modest goals—e.g. to describe idealized phenomena in limited domains. If anything, his religious commitment within the Catholic Church increased with time.[9]

15.3 Cosmology and widening consciousness

It is notable that the progress in astronomy and cosmology over the past few hundred years has tended to make it more and more apparent that planet Earth is not central, in either the Solar System or the galaxy, or the wider universe. This is not to deny that a planet with abundant life on it remains remarkable. Earth and its inhabitants certainly merit a large fraction of our scientific and other attention, as compared with

---

[9] John Heilbron, *Galileo* (Oxford University Press, 2012); a readable non-specialist account is Dava Sobel, *Galileo's Daughter* (Fourth Estate, 1999).

planets with little or no life. But it remains interesting to know that Earth is just a small part of a larger picture, and many of us find this rather a welcome than an unwelcome thing to know. This knowledge seems to allow our consciousness to expand a little, and our self-importance to diminish; both are welcome outcomes.

In this area we would like to suggest that a certain 'just-so-story' or simplistic piece of historical analysis should be held up to critical scrutiny.

It is commonly asserted (in modern-day cosmology textbooks, for example) that the reason some people didn't like Copernican astronomy is not because they thought the data indicated that the Earth stayed still, but because of other and less worthy motivations. It is claimed that 'religion' or 'theology' taught that Earth was central and important, and therefore Earth must be located in a physically central place in the cosmos. It is claimed that such thinking promoted a restricted world view, and held back the kind of expansion of consciousness which now we enjoy.

This 'just-so-story' is probably fair comment on a few individuals with confused minds back in the seventeenth century. But this does not give us the right to assert it as a generalization, because that would be unfair and unjust. One must be especially careful not to unjustly accuse people who are not present to defend themselves. So, we would like to invite a competent historical enquiry into this question.

To motivate such an enquiry, let us note a couple of relevant points.

First, in pre-Copernican Europe, the location of Earth was commonly associated not with importance but with lowliness. The Earth was the place of fault and failure; the heavenly bodies (Sun, stars, other planets) were associated with purity and perfection. This is why it was equally remarkable, to people at the time, not just that the Earth should move, but also that the Sun was itself not of one pure whiteness, but had spots. So, when Earth was displaced, as it were, from its central position, it was not moved to a more humble place in their thinking; it was moved towards a role that previously was reserved for the 'more perfect' bodies. The result is that the rearrangement did not point to a humbler role for Earth, but merely to a different arrangement of positions and motions. The change was neutral as regards how it could be interpreted as a metaphor for importance.[10]

---

[10] The following analogy may help to get the flavour of this. When people in the past thought of the Earth as in the middle, it was mainly because that is how it looks and this

The second relevant point is that the centres of learning in Europe (monasteries, schools, and universities) had enjoyed very large support from theologically motivated efforts over many centuries, and established churches were among the chief patrons of astronomy. In view of this, the overall impact of theology on this particular development in science cannot be quite as simple as the just-so-story would have us believe.

What would be welcome now is a chance to get a more informed and fairer appreciation of all this. That is the type of expansion in our consciousness that we could benefit from.

15.4 Summary

To return to the original question of the chapter: what's a rational person supposed to do if they find their religious beliefs in conflict with the claims of science? As should be clear from the preceding, there's no simple answer to this question. We cannot counsel that every scientific 'fact' should automatically be taken as hard data to which religious belief must be adapted. For if we gave that counsel to our thirteenth-century thinkers, then we would have them all conceding that the universe is eternal, in the sense of going on and on into past time—in sad conflict with what most physicists now believe is true. But nor can we counsel that the religious believer retrench and claim that the science must be wrong if it conflicts with religious belief. That approach is flawed primarily in how it would retard the free growth of science. Just think again of the story of Bonaventure. While Bonaventure might have reached the right conclusion—that the universe is *not* eternal—he reached it for the wrong reasons. Indeed, if Bonaventure, rather than Aquinas, had held sway over the future of Western thought, then science might very well not have flourished in the way that it has. After all, there is good reason to think that Aristotelian physics—while mostly wrong—was an important stepping stone toward modern physics.

Thus, we agree with the convergent realist picture that a false scientific theory can serve as a stepping stone to a better scientific theory. But the progress of science is most definitely not a unidirectional progression

hardly deserved a second thought. But being in the middle was not necessarily felt to be a welcome place. Being the centre of attention is welcome to a performance artist, but unwelcome to an accused person standing in the dock in a court of law. It all depends on how you see yourself.

toward some ideal truth of the matter. That would be far too simplistic. It would be more accurate to say that science progresses by pushing us to adopt a broader and broader perspective. As we saw with the case of Galileo, physics doesn't show that geocentrism is wrong—rather, it shows that abandoning the presupposition of absolute rotation leads to a more powerful way of describing and systematizing phenomena.

One might be tempted here to say that there is still a sense in which science is progressing closer and closer to the truth. For one might think that the claim 'there is no preferred frame of reference' is closer to the truth than the claim 'there is a preferred frame of reference'. But things aren't so simple. What is it supposed to mean to say that 'there is no preferred frame of reference'? For a frame of reference is not a thing like a rock or a planet. To say that there is no preferred frame of reference is not a simple denial of existence, as it would be to say that Santa Claus doesn't exist. Instead, to say that there is no preferred frame of reference is sort of like enunciating a rule: *we can describe the phenomena, with equal validity, from the point of view of many different frames of reference.* But now we're talking about what we 'can' do and about 'validity' of descriptions and other things that have no direct analogue in physical reality. In short, to say that 'science teaches that there is no preferred frame of reference' would be something of a solecism. It is much more accurate to say that it has proven fruitful not to restrict ourselves to a single frame of reference.

So, all this talk about religious belief being 'static' and scientific belief being 'dynamic' and 'moving ever closer to the truth'—it's all based on shaky metaphors and misleading pictures, not to speak of a basic misunderstanding of how religious commitments function in a person's life. A better way of speaking would have us say the following sorts of things.

On the one hand, a religious outlook involves a complex of attitudes, motivations, and commitments, among other things. Some religious commitments may have a cognitive aspect; i.e. they might involve commitment to the truth of certain claims. For example, a religious person might be committed to the claim that 'God created the universe' or more controversially, to the claim that 'while we were yet sinners, Christ died for us'.[11] However, these sorts of claims are rarely intended to function as simple reports of fact, as, say, the claim that there is a

---

[11] Romans 5:8.

lamp on my desk. Religious claims typically carry implicit moral content. For example, when a person says, 'God created the universe', she is typically signifying not only a factual claim, but also an attitude—typically, that there is a being of high moral stature who deserves our respect and obedience. (If she didn't have this attitude, then she'd probably qualify her statement by saying that some being created the universe, but that being isn't God in the traditional sense of the word.)

Similarly, we (the authors) believe that some events in our own lives are most accurately described in explicitly religious language. For example, someone might say that 'God met me in prayer this morning'—and they wouldn't mean it as some kind of loose metaphor, e.g. as indicating that they had a particular sort of psychological experience. They would intend their words to be taken nearly at face value, as the best way they know how to describe the situation and their attitude towards it. Describing such events in the impersonal language of the exact sciences would actually detract from our ability to understand them.

Let's recall how this chapter started—with the 'triumphal march of science' picture. According to that picture, once upon a time, people believed what religion said about the universe and our place in it, e.g. 'the universe was created by God.' But then along came science, and it said: 'No, those claims are false; what really happened can be explained in purely materialistic terms.' But that story is far too simplistic. Rather, the practice of science calls us to adopt new standpoints, and new vocabularies to go along with those new standpoints. And we think that these standpoints can enrich a religious perspective. Learning to speak a new language, and committing oneself to certain claims in that language, is perfectly compatible with maintaining commitments that one only knows how to express in a different language.

16

Miracles and reasonable belief

Hans Halvorson recounts:

Two days after my son was born, he began to have trouble breathing. We took him to the paediatrician, who was puzzled by our descriptions of the problem. But then, while we were in the paediatrician's office, our son's face began turning a deep red, and then purplish. The paediatrician, in an alarmed tone, told us that we should take our son straight to the emergency room at the hospital. So, we did.

For five days, my wife and I stayed at the hospital with our son. He was attached to a pulse-oximeter, and his oxygen levels were dangerously low. And yet, doctors had no diagnosis—they didn't know what was wrong with him. We could do nothing but wait, and prepare for the worst.

Then on the sixth day, his oxygen levels stabilized and the colour began to return to his face. He opened his eyes and seemed to regain his energy and hunger. The doctors didn't know what was going on. All they knew was that he seemed now to be a perfectly healthy baby, and there was no reason to keep him in the hospital any longer. Nervous as we were, we took our son home.

Today is my son's tenth birthday. He is a strong, healthy boy, with no serious health problems. Many Christians would say that his life is a miracle. But then what about that family sitting beside us whose son died the day we left the hospital? When I pause to think about it, I have at least four close friends who have lost children. One of them went through an experience almost identical to mine, only his baby didn't survive. In light of these facts, I'm loath to say that God stepped in to save my son—as if somehow God preferred him over the children of my friends.

During my son's time in the NICU, many people were praying for him: myself, my wife, our parents, our siblings, our church, etc. I don't know what each of them prayed. I'm sure that some prayed for healing, and others prayed that my wife and I would have peace. I suspect that some even prayed for a miracle. I myself didn't pray for a miracle, because—truth be told—I don't know what I would be asking for. I did pray that

God would give this boy a chance to grow, and to learn about God just as I had when I was a boy. I didn't suggest any particular method of healing to God, because surely He knows best.

Despite our sense of absolute dependence on God, we did everything within our human power to get the best medical attention possible for our son. We drove over an hour to the nearest major city in order to get a higher level of medical care. We urged doctors to perform additional tests in order to rule out possible problems. We spent more time trying to find 'naturalistic' solutions to our son's problems than we did praying. But we also prayed. The two activities—praying and working—are not mutually exclusive. Quite to the contrary, praying fortified us for the battle ahead; and watching those brilliant doctors filled us with a sense of thankfulness to God.

We still don't know what was wrong with our son, and we still don't know how he was cured. Perhaps everything that happened could be explained in terms of natural causes and effects. Or perhaps it can't, perhaps a miracle occurred. I don't know—and, honestly, it doesn't really matter to me what mechanisms were involved in my son's apparently miraculous recovery. My thankfulness to God is the same, no matter how the healing came about.

As you can see from what I've just said, I'm not convinced that miracles never occur. For all that I know, a miracle occurred in the NICU in February 2007. But it would be beyond presumptuous of me to go around proclaiming confidently that a miracle did occur. How would I know what God was doing, and how He did it? Am I so sure that I know the laws of nature, and that they were broken in that intensive care unit? Of course not. A fair assessment of the situation is that, despite the element of mystery, there isn't a shred of definitive evidence that a miracle occurred.

16.1 Can scientists believe in miracles?

Hans' reaction, as just described, is typical of the attitude that we (the authors) bring to the question of whether miracles have occurred, and of whether they continue to occur. This is the kind of area where we have learned to live with unanswered questions. On the one hand, we believe that God has acted, and continues to act, in the course of history and in our own lives. On the other hand, we believe that miracles have only a small role to play in moving a person to become a follower of Jesus. In what follows, we will begin by asking what it means to say that an event is a miracle. Then we will discuss various arguments against the possibility of miracles, and against the possibility of rational belief in miracles. We find most of these arguments to be wanting. The one argument that we find convincing is an argument driven by our Christian commitment to the scientific enterprise: we should see surprising events as a challenge for scientific explanation rather than as an excuse for scientific laziness.

16.2 Defining miracles

In the Western world, philosophers have always been concerned with the meanings of words. For example, Socrates spent a lot of time talking about the meaning of words such as 'justice', 'knowledge', and 'goodness'. In the twentieth century, this concern with the meaning became even more central in academic philosophy. In the English-speaking world, philosophers' primary goal was to figure out what words meant. And lest you think that these sorts of investigations are idle curiosities, just think about the impact that definitions can have on public policy decisions. For example, in a 2005 trial in Pennsylvania (*Kitzmiller v Dover*), a judge ruled that Intelligent Design Theory could not be taught in public schools because it is not science—a judgement which assumes, of course, some particular definition of the word 'science'. Similarly, if somebody says that it's irrational to believe in miracles, then a lot hangs on their definition of 'miracle'.

We've come to think that it's very difficult, if not impossible, to come up with a good, general definition of 'miracle'. And in the absence of a definition of 'miracle', it's not possible to answer the question, 'Can it be rational to believe in miracles?' Certainly, we can ask more specific questions, such as 'Can it be rational to believe that Jesus rose from the dead?', but we cannot give a blanket answer to all questions of that sort.

But don't take our word for it—try yourself to come up with a good definition of 'miracle'. For example, suppose that you begin with the following proposal:

An event is miraculous if it violates the laws of nature.

This definition has a number of problems. First and foremost, when a scientist sees a 'violation' of the laws of nature, then she shouldn't conclude that a miracle has occurred. Indeed, we teach our students to do the following if an experimental outcome runs contrary to the known laws of nature: first of all, run the experiment again to see if something was wrong with the setup. And only after the experiment is repeated several times should the student entertain the thought that perhaps we were wrong about the laws of nature. In that case, we still don't consider that a miracle occurred. Instead, we think we are learning something about the laws of nature which was previously unknown or misunderstood. (We even hope that one of our students might discover such a thing, for that would be a genuine scientific advance.)

The definition we've suggested is also unsatisfying because it defines one murky concept, 'miraculous', in terms of another murky concept, 'laws of nature'. Philosophers are already well aware that it's very difficult to find a good definition of 'laws of nature', and some have gone so far as to conclude that there are no laws of nature. (Their reasoning goes roughly as follows: a law of nature would have to have a bunch of different features, but these features are mutually incompatible. Therefore, there are no laws of nature.) While we can't get deeply into this issue, let's consider one obvious proposal:

A law of nature is an exceptionless regularity.

But this definition won't work, because there are exceptionless regularities that aren't laws of nature. For example, the statement

Every black US president was born in Hawaii

is an exceptionless regularity, but we have strong intuitions that it's not a law—based on the fact that it *could be* violated. So, a law of nature is more than just an exceptionless regularity—it's a regularity that *could not* have an exception.

This game—of proposing definitions, checking them against examples, and modifying the definition if necessary—could go on and on, as it does on page after page in academic philosophy journals. We're not trying to say that it's impossible to find a good definition for 'law of nature'. (In fact, scientists seem to do just fine with their intuitive concept of laws of nature—as general principles that organize the phenomena they investigate.) Our point is simply that it's not totally obvious what's meant by 'law of nature', and so if a miracle is a violation of the laws of nature, then it's not totally clear what miracles are either.

Even if we had such a definition of 'miracles', it wouldn't help us understand the meaning of the miracle stories of the Bible. The problem is that the authors of those stories didn't themselves have a precise definition of the miraculous, and so their intention couldn't possibly have been to assert that a miracle—in this precise sense—did occur. No, the biblical authors don't generally speak directly about metaphysical issues, and they aren't aiming at scientific precision. They tell us that certain events happened, and the significance of those events. They don't say anything about whether those events have some sort of naturalistic explanation.

The philosopher Richard Purtill recognizes the difficulty with defining miracles in terms of laws of nature: while the concept of a law of nature wasn't clearly articulated until the seventeenth century, people in biblical times recognized events as miracles.[1] Perhaps we should define miracles as events that deviate from the natural course of things, where the concept appeals to common sense rather than to tutored scientific knowledge of nature. This looser, more commonsensical definition has the advantage that biblical people could recognize an event as a miracle. However, it also has the disadvantage that it presumes some relatively stable notion of the 'normal course of events'. Unfortunately, it's simply not the case that this notion has been stable, not even over the past few hundred years. For example, is it in the normal course of things for iron fillings to spontaneously align themselves? Of course not—that's why it seemed so strange to nineteenth-century scientists like Michael Faraday.

To take another example: imagine that you grew up in the village of Covington described in the 2004 movie *The Village*.[2] Suppose that you are out walking your dog, and you find a shiny metallic rectangular object on the ground. Just when you pick it up, your dog—incidentally named 'Siri'—starts to wander off into the woods. So, you yell out, 'hey Siri, where are you going?' At that moment, the square black thing lights up, and a voice speaks, 'Hello Rachael, I'm not going anywhere.' In this case you might, with good reason, think that you've witnessed

---

[1] Richard L. Purtill, 'Defining Miracles', in R. Douglas Geivett and Gary R. Habermas (eds), *In Defense of Miracles: A Comprehensive Case for God's Action in History* (InterVarsity Press, 1997). As Purtill is well aware, this description is somewhat over-simplified. First of all, even if ancient peoples didn't have an exact concept of law of nature, might not they have possessed some primitive version of that concept— a version sufficient to grasp the sense in which miracles violate laws? Second, there is some evidence that some ancient peoples *did* have the concept of law of nature, e.g. the Stoic philosophers.

[2] The movie is set in the present day, but the residents of Covington believe that it's still the nineteenth century. In particular, they have never interacted with the technologies developed in the twentieth and twenty-first centuries.

something miraculous. But those of us on the outside know better; we know that it's just an iPhone.

As a final example, consider the question of which motions are natural or unnatural. This question was of central concern for the philosopher Aristotle. According to Aristotle, all objects have some particular home in the universe, and they are constantly striving to return to their home. Thus, for Aristotle, a natural motion of an object is a motion towards that object's home. For example, if I drop a rock, it naturally falls to the ground because earth (and rocks) belong in the centre of the universe. Similarly, if I blow a bubble underwater, it naturally rises to the surface, because air belongs upward, above earth and water.

In contrast to natural motions, Aristotle says that a motion is unnatural or 'violent' if the object moves further away from its home. Thus, for example, throwing a rock in the air is an unnatural motion for the rock.

Given Aristotle's notion of 'natural' motions, it makes sense that what needs to be explained are the non-natural motions. For example, if you see a rock falling to the ground, then there's nothing particularly strange about it. But if you see a rock rising in the air, then something weird is going on, and it demands an explanation.

Aristotle's physics dominated the West until roughly the time of Galileo. At that time, Galileo made the audacious suggestion that all motion is relative. In other words, there is no sense in which particular objects are stationary, and other objects are moving. Instead, the only interesting question is which objects are moving relative to each other. And thus, there is no sense in which it's more natural for an object to sit in one place than it is to move—despite what our ancestors may have thought. Accordingly, it's not clear which events are ordinary, and which are extraordinary.

You might now be wondering whether our view of miracles could possibly be an orthodox view. If we don't know how to define 'miracle', then how can we possibly affirm that 'miracles are possible'?

We don't need a precise definition of 'miracle' in order to grant credence to the remarkable biblical stories—such as the Israelites crossing the Red Sea or reed sea, Jesus being conceived without a human father, Jesus turning water into wine, and Jesus raising Lazarus from the dead, for example. We can say 'miracles have occurred', but in saying this, we're simply affirming our willingness to take these stories on their own terms. Yes, some sophisticated theologians might think that our view is rather naïve. Unlike those theologians, we don't have much interest in the project of 'demythologization'. We're not going to try to sort the biblical stories into those that are naturalistically acceptable, and those that are not. We take all of these stories quite seriously as conveying truths that are important for growth in wisdom, and often more personally significant than even our best scientific theories.

These biblical stories might be considered to be implausible from the point of view of contemporary science. Why then do we think that they occurred? In short, our starting point is Jesus' call to follow him. He says to us: here is your condition, and here is my remedy for it. We recognize that he has described our condition accurately—more accurately than anyone else. And he promises us things—such as changed lives—that we know we need. Thus, recognizing that Jesus has spoken the most profound of truths, we trust other things that he says—e.g. when he says that after three days he will rise again.

In this, we have adopted as a first attitude the decision to be receptive to what people showing signs of honesty and care about reporting have told us. We are referring to the groups of people who wrote the New Testament documents. This does not mean necessarily accepting everything at face value on first reading; rather it is by this process, combined with wider concerns and links to people in the present, that one may come to judge that in Jesus one is encountering a uniquely significant person. One then becomes more receptive to the notion that his insight into what is and is not possible may be, in important areas, greater than one's own. Of course, it is possible that when John, or the community around him, wrote that Jesus turned water into wine, what really happened was something else entirely, and the writer is confused or has misinterpreted what happened. If this incident had been inconsistent with what we otherwise know about Jesus, then that would be, most likely, the correct way to account for it. But this was not an isolated incident, so we should explore further before leaping to such a judgement. After exploring further, one may come to the conclusion that this was a judiciously chosen moment in which Jesus decided to teach an important lesson, and many gallons of water became wine on that occasion.

If you ask us, how exactly did those events happen? How, for example, did Jesus turn water into wine? At this point, we demur.

We don't know exactly how it happened. How would we know that? We weren't there, and the eyewitnesses didn't provide any further information. We don't know if God did it by means of cleverly orchestrating purely natural processes (whatever that means), or whether God decided to introduce His creativity more directly. We don't know, and you don't know. But not knowing how exactly it happened does *not* mean that it's irrational to believe that it did happen. The issue of rationality here is more to do with how we assess the community that produced the account: what we think they were doing and whether their report can be trusted. This, in turn, connects to the wider issue of what was taking place in first-century Palestine and in the Jewish history that led up to it. We are drawn inexorably to the issue that these miracle stories fit into the narrative of the Incarnation. We take that narrative seriously. We cannot adopt a blanket denial of all the accounts of miracles associated with Jesus' ministry. Such a denial amounts to taking the following attitude: our ideas about what is possible are so final that they can and should prevent us from allowing that God can inaugurate something genuinely new in human history. Our sense of intellectual duty constrains us to be open to that possibility of novelty.

But, and this is a key point, we don't take the Incarnation narrative as data that must be explained by our best scientific theory of the world. By that story's own lights, the events that took place were *historically singular*, and are deeply imbued with spiritual significance. Thus, we simply do not know whether those events (such as water turning to wine) are things that could happen in the 'ordinary course of nature' when some particular circumstances are extant. What we do know, based on the biblical authors' testimony, is that God was directly active in bringing about that particular event. Thus, there is prima facie reason for thinking that that event is an outlier. We're not saying that the biblical events are purely spiritual, and have no connective to the historical narrative. The question isn't whether the Bible is true or false—it's about what it's trying to teach us.

Lacking a general concept of the miraculous, what's our position on the possibility of miraculous events today? If pressed for an opinion on that question, we have a simple answer: there is no convincing argument which shows that God could not perform miracles today. In that simple sense, we affirm the ongoing possibility of miracles.

16.3 Arguments against miracles

Our view on miracles might be considered a decision to avoid two extreme views. The first extreme view is the simple 'no miracles' view. The second extreme view is the simple 'yes miracles' view. Let's first describe these points of view. To be clear, we're not trying to describe the view of any particular person. What we're trying to do is to carica-ture two extreme views that might bear some resemblance to views that some people may hold. To our minds, each of these views gets something right, and both of these views get something wrong.

There are some people who think either that miracles aren't possible or that even if they are possible, it would still be irrational to believe that a miracle has occurred. Why would anyone think that miracles are simply impossible? There are a couple of ways that miracles could be simply defined out of existence. First, if one supposed that a law of nature is (among other things) an exceptionless regularity and that a miracle is a violation of a law of nature, then by definition miracles would not occur. That's way too fast and completely uninformative. We need to do better than that.

Another way one could define miracles out of existence is by saying that a miracle is a supernatural event. But all of human experience is, by definition, human experience. And humans live in the natural world. So, by definition, everything we experience is natural. Therefore, we cannot experience miracles.

It's a bit more difficult to see what's wrong with that simple argu-ment. I'm sure some defenders of miracles would simply object to the claim that all of our experiences are natural. Some defenders of miracles might claim that humans could experience something beyond the natural. But let's be clear. Even if a human experienced something beyond the natural, the experience would be mediated by natural things. For example, to see a miracle, light rays must impinge on a person's retina, etc. What this means is that it's not possible to experi-ence the supernatural directly, but only indirectly by means of our embodied experience.[3]

Perhaps we've gone too far. Some of you might think that human beings have immaterial souls, and that our immaterial souls could

---

[3] A confession: we don't know how to draw the natural–supernatural distinction, except to say that God is not a natural thing, and everything He created is a natural thing.

directly experience other immaterial realities. Suppose for a moment that we don't dispute such a claim. Nonetheless, that claim is of no help at all in understanding claims about miracles. For claims about miracles are all claims about something happening in the natural world—even if it's not something that would ordinarily happen. For example, water turning into wine is not a direct spiritual experience— it's a change from one physical state to another physical state. When Christians say that Jesus changed water into wine, they are saying that a person at the wedding in Cana could have seen that the liquid was originally water, and then could have seen that the liquid was subsequently wine. There is no further claim that the observers of that event experienced some non-natural thing. They may very well have understood the supernatural significance of the events they experienced with their natural sense organs. But their access to those events was by means of their ordinary five senses.

In other words, to grant credence to biblical miracle stories is not to claim that humans experienced something that humans could not possibly experience, much less describe to other people. No, these stories talk about relatively shocking events in the natural world that have some profound spiritual significance. Whether those events in the natural world occurred cannot be ruled out by definition.

What follows from the preceding discussion is that the possibility of miracles cannot be ruled out by the very definition of the word 'miracle'. Or, if it could be, then we could simply change the definition of what a miracle is.

16.4 Miracles and irrationality

Some people think that even if a miracle happened, we couldn't rationally believe that it did. We disagree. We already said that we consider it to be true that some miracles have occurred; e.g. Jesus rose from the dead. And we wouldn't believe that Jesus rose if we thought it irrational so to believe. Perhaps then we need to be a bit more clear about what it means to be 'irrational'.

Typically, the word 'irrational' connotes a culpable failure to use the tools of human reason. For example, Holocaust deniers are irrational because they don't adjust their beliefs in proportion to the evidence that they have at hand. In that case, we certainly don't condone irrationality— which would be tantamount to condoning a sort of moral failure.

Nonetheless, we will explain later a sense in which believing in miracles might be seen as 'irrational', although not in any blameworthy sense. Some theologians would prefer that we use the word 'a-rational' or 'supra-rational' instead of 'irrational'. These theologians would say that Christian belief transcends reason, but does not conflict with it. We think that there is something right about that point of view, but that this notion of 'transcending' can be further clarified. We'll return to that point, but let's first think about why somebody might believe that it's irrational, in a morally blameworthy sense, to believe in miracles.

A person might define 'miracle' so that it's automatically irrational to believe in miracles. For example, suppose that we define miracles to be events that could not possibly occur. Then no rational person would believe that miracles could occur.

A weaker definition of miracles would suffice for an argument of this sort. If a miracle is defined to be an event whose occurrence is extremely unlikely, then it wouldn't be rational to expect a miracle.[4] And even if you thought you experienced a miracle, the most rational thing to do might be to question the veracity of your senses, or your sanity. What's more likely, that a man walked on water, or that you mixed up your memory with an overactive imagination? Furthermore, if miracles are, by definition, unlikely to occur, then you should suspect that anyone reporting a miracle is probably either deluded or lying.

But that would prove too much—because we do sometimes accept testimony that some unlikely event occurred. Suppose that Professor Johnston from the office next door runs over to my office and says, 'The lukewarm cup of tea on my desk just spontaneously boiled!' Now I have a dilemma. On the one hand, I believe that such events are extremely unlikely. On the other hand, here is one of my esteemed colleagues, reporting that he witnessed such an event. What would a rational person do?

The eighteenth-century philosopher David Hume argued that a rational person would never believe that a miracle occurred. In particular, his argument says that a rational person in my shoes would necessarily disbelieve the testimony of his colleague. Hume's point is not about whether miracles can occur at all; it is about the possible

---

[4] By definition, you expect those things that you believe are likely to occur. Certainly, you can hope that an unlikely event will occur—but hoping is not expecting.

unreliability of reports or evidence about extraordinary events. He argues that the balance of probability must always favour the conclusion that a reported extraordinary event did not happen.

Hume helpfully focusses attention on the issue of evidence and reliability, and this is a useful progress. However, his argument has failed to convince many philosophers, and it has failed to convince us as well. Why should I believe that it's unlikely for a cup of tea to spontaneously boil? Hume describes the situation as if I had observed many cups of tea, and had never seen one spontaneously boil. But the fact is, I haven't observed that many cups of tea. Perhaps a thousand? And for how many combined minutes? I really haven't observed enough tea on my own to draw any firm conclusions about its behaviour.

Nonetheless, I am a member of an intellectual community of tea drinkers. And if we assemble all of our experience together, then it forms a sample of reasonable size. In order for this data to count as data for me, I must trust my colleagues and peers that they have reported the data accurately. Thus, I shouldn't necessarily immediately dismiss my colleague's report, for the evidence against his report is an aggregate of reports from yet other colleagues. Humans are faced, quite generally, with the problem of assembling reports from a diverse array of witnesses. Which witnesses are to be trusted, and which are not to be trusted?

It would be bad if we always rejected the testimony of contrarian witnesses. Just imagine several people watching a football match, say the famous match between Argentina and England in the 1986 World Cup. Most witnesses say that Diego Maradona headed the ball over Peter Shilton and into the goal. But a couple of witnesses say: No, we saw his hand come up and hit the ball over Shilton's head. Should we reject the testimony of the latter witnesses simply because they are in the minority? Not necessarily. We may want to take their report with a grain of salt—perhaps these are English fans who have a vested interest in their team winning. Nonetheless, even if most people disagree with their report, we don't want to blot their report from our register, so to speak. Their report is further data that must be incorporated into an adequate account of the situation.

Let us add some stronger examples. In 1610 it was a universally and very frequently observed regularity of nature that all astronomical bodies orbited the Earth. So, according to Hume's argument, the rational reaction, on hearing reports that there were moons orbiting

Jupiter, would be to dismiss the reports, on the argument that it is more likely for the observer to have been mistaken than for the well-established regularity of nature to be broken.

In a famous physics experiment carried out by Hans Geiger and Ernest Marsden under the direction of Ernest Rutherford in 1910, scintillations were seen on a screen, which implied that particles with a large momentum (on an atomic scale) could be reflected by atoms. In order to understand how surprising this was, we can listen to Rutherford's reaction: 'It was quite the most incredible event that has ever happened to me in my life. It was almost as incredible as if you fired a 15-inch shell at a piece of tissue paper and it came back and hit you.'[5]

In a final and tragic example, it has often happened that attempts to help primitive peoples with medical aid have been resisted because the people could not conceive of how the types of treatment on offer could possibly be better than the traditions of their community, and the world view accompanying it.

In all these examples, the best response to the strange report is not to come to an immediate definitive decision. One may well judge, at the outset, that it is more likely that the observer was mistaken than that the event occurred. However, one does not stop there. The best response is to *explore further*. In each case, what one is in search of is evidence either that the report was indeed mistaken or that one can receive the new idea because it is part of *a coherent larger view* of the nature of reality.

We've just been talking about Hume's argument against the claim that a miracle occurred. But note that this very claim contains an ambiguity. Are we asking whether miracles in general can occur? Or are we asking whether some particular event (which some people might classify as a miracle) has occurred? In the former case, we might think that miracles are (by definition) something that is unlikely to occur. In the latter case, different considerations come into play. Hume's probabilistic argument against miracles is a gross oversimplification. Hume made some great points about the relationship of evidence to theory—at least for someone living in the eigteenth century, struggling to understand what probability means. But from our contemporary perspective, the limitations of Hume's argument are apparent.

[5] E. N. da C. Andrade, *Rutherford and the Nature of the Atom* (Peter Smith Publishers, 1964), 111, and Laylin K. James, *Nobel Laureates in Chemistry, 1901–1992* (American Chemical Society, 1993), 75.

Consider the most important miracle of the Christian religion, the resurrection of Jesus from the dead. Is this an improbable event, the man named Jesus rising from the dead, in that particular place and at that particular time? The answer to this question is not as simple as it may seem.

Every event belongs to many different reference classes. Consider, for example, the following event: I just accidentally knocked a pen off my desk. Was that event improbable? It depends on how you describe it. All of the following are correct descriptions of this event.

- A physical object falls approximately three feet to the ground.
- A pen falls approximately three feet to the ground.
- A pen falls from a desk to the ground.
- A pen falls from Hans Halvorson's desk to the ground.
- Hans Halvorson's favourite pen falls from his desk to the ground.
- Hans Halvorson accidentally knocks a blue pen, with no cap, off his desk, and it then falls to the ground at approximately an 86° angle to the floor.

Each of these descriptions gives rise to different expectations from the others. For example, suppose that I usually have only black pens on my desk, but because of a bunch of random coincidences, today I happened to have a blue pen on my desk. Then the event as described in the last sentence has a much lower probability than the event as described in the previous sentences.

In the above example, the more specific case has a lower probability than the less specific case. However, it can also happen the other way around. Take, for example, the probability that someone drawn from the human population might gain an average score of close to 100 in Test Cricket, over a career of at least 20 innings. The probability of this is perhaps around one in a billion or one in ten billion. But if we add the specific information that that person is Don Bradman, acclaimed Australian cricketer, then the probability is (for all practical purposes) one.

Now let's apply this reasoning back to the resurrection of Jesus. The description 'a man rises from the dead' is just one of the possible ways of describing the actual event (which transcends any particular description of it). Under this description, the event of Jesus' resurrection has an admittedly extremely low probability, although it's still non-zero, because there is no strict logical impossibility involved. Now, if the probability of 'a man is raised from death' is not zero, then the probability that some particular man, say Jesus of Nazareth, rises from the dead might actually be quite high. That's how it works with statistical laws: while the typical instance might have a low probability, some specific instance might have a much higher probability.

Let us reiterate at this stage, however, that we do not proceed in a gullible or gung-ho manner when it comes to miracles. We are under a moral duty to be wise, and wisdom includes that we should be aware that human reports of human experience can be mistaken or misleading. In view of this, if the New Testament accounts of meetings with Jesus after his death were all the evidence we had to go on, then the weight of evidence would overwhelmingly favour the view that the accounts somehow came to be written as they are even though no resurrection took place. However, when we bring in the evidence of the whole of Jesus' life and of what came both before and after (the experience of the history of Israel, and the experience of the Christian community), then, in the opinion of many reasonable people at least, the weight of evidence is balanced. One cannot be sure either way, and it is not irrational to explore further in the direction of 'yes, this happened', just as it is not irrational to explore further in the direction of 'no it did not'.

There is one last gasp in Hume's argument against miracles: perhaps there is a *privileged* description of each event that yields its objective probability. We (the authors) are open to this possibility. But where should we expect then to find such a privileged description? Would it come from 'folk wisdom', which tells us (among other things) that cups of tea cannot spontaneously boil? Well, folk wisdom also tells us that humans cannot fly in the sky—a piece of wisdom long ago falsified, when the Wright brothers invented the airplane. It seems that if there is any authority on whether physical events are possible, then that authority is found in *fundamental physics*, not in folk wisdom. However, fundamental physics doesn't traffic in concepts like 'persons' or 'resurrection'. To get from the objective probabilities (if there be such) of fundamental physics to facts about what's possible in everyday life, we must extrapolate, approximate, and guess. In other words, even if there are objective probabilities at the fundamental level, they don't directly answer the question of whether a person who gave himself unreservedly into God's keeping can rise from the dead. We must be very careful in talking about the probability of events in the past. There are all sorts of things that have happened in each of our lives which might have seemed extremely improbable beforehand. But they did happen, and that is that.

16.5 Pragmatic arguments against miracles

There's one more argument in the arsenal against miracles. This argument, we think, has something right about it. But what's right about it doesn't tell us that miracles don't happen. It just reminds us that human interaction with the world is active, not passive. The world doesn't just come and tell us what it's like. To the contrary, we must actively pose questions to the world.

The final argument goes like this:

> *Believing in miracles would have bad consequences. Therefore, you ought not to believe in miracles.*

What are these supposedly bad consequences? Two of the main ones are (1) hampering the progress of science, and (2) not searching diligently for natural solutions to life's problems. As to the second consequence, we are thinking in particular of medical science, and the worry is that belief in miracles might undermine the search for legitimate natural solutions to medical problems, or might inhibit an individual from seeking medical assistance when they need it.

Let's turn to the first supposed consequence: if you think that miracles can happen, then you cannot do science. A standard refutation of this claim is to point to examples of top-rate scientists who nonetheless believed in miracles. The list here is long: Boyle, Newton, Maxwell, Stokes, Compton, LeMaitre, etc. But that rebuttal has not proven to be convincing. Why not? Because the claim 'if you believe in miracles, then you cannot do science' wasn't really intended to be a description of what actually does occur. Consider an analogy: as a teacher of logic, Hans tell students that they should not believe contradictory things. He might even be heard to say: 'You cannot believe contradictory things.' But of course, they *can* believe contradictory things, and probably many of them do—despite their professor's urgings to the contrary. In the same way, the claim 'you cannot believe in miracles and be a good scientist' is meant as a normative statement—it's saying that you *ought not* simultaneously adopt the beliefs necessary for doing science, and also consider it to be true that miracles can occur.

Let's see if the normative claim—one *ought* not to entertain the possibility of miracles while doing science—is plausible. First thing: what are these 'beliefs necessary for doing science'? Are there any such things? We think that there are, though others would contest that claim.

It may be helpful to think of science as a two-step process. In the first step, the scientist adopts certain rules for the game. These rules tell us what the goal of science is, and what standards we'll hold ourselves to in playing the game. In the second step, we go out and play the game known as scientific investigation. Of course, real science is not so simple; these two steps are never truly distinct. The practising scientist is constantly renegotiating the rules by which the game of science is played. Nonetheless, it can be helpful to 'resolve' scientific practice into these two components, rather as forces in physics can be resolved into vectors in different directions.

The next question is whether the rules of science permit belief in miracles. But that question is ill-formed. For the rules of science aren't about what you can or cannot believe. If that were the case, then science would truly be dogmatic. No, the rules of science are about good procedure—they are rules like 'state your hypothesis clearly', and 'run multiple trials to make sure that you result is reproducible'. Those procedural rules don't tell you that you shouldn't believe in miracles.

And yet, there is something about the rules of science that does relate to miracles. The rules of science essentially tell us that miracles are not interesting to science. The rules of science tell us that we should use clear language, not emotively charged language. But to describe an event as a miracle is about as emotively charged as anything can be— witness the hostile reactions to miracle claims. Similarly, the rules of science push us toward reproducibility, again something we cannot possible expect to hold of miracles. Science wants to describe things that we humans can manipulate, manage, systematize, and describe. If miracles occur, they aren't the kind of event that science concerns itself with.

But now, suppose that you accept the rules of science. Can you believe in miracles? Well, there's another important question we need to ask. Does accepting the rules of science mean accepting them as a sort of religious creed that will guide all aspects of your life? What a question! Just imagine treating your spouse, or your children, or your best friend, as if they were objects in your laboratory. Imagine applying the same rules that you apply to scientific hypotheses to claims that your spouse makes—e.g. the claim that he or she loves you.

Consider the point from another angle. Suppose that we were to design robots in our own image, and we were programming these robots to behave perfectly rationally. You might think that we should tell the robots to be perfectly scientific in each and every interaction. But then these robots would have horrible interpersonal relationships!

No, if we were to design robots in our image, then we ought to give them different 'modes' and allow them to switch between these modes. We should give them an 'investigation mode' in which they formulate hypotheses and proceed to test them. And we should give them a 'relational mode' in which they attempt to empathize with other conscious beings.

One thing that is amazing about human beings is that our modes of being are not binary, or even discrete. It's not as if we have 'science mode' and 'personal interaction' mode, and nothing in between. It's also not the case that the two modes are mutually exclusive. When we relate to other human beings, we cannot completely turn off the scientist mode—we are constantly making up little hypotheses and putting them to the test. But to think that all rational thought is of one kind, that strikes us as ridiculously oversimplified and phenomenologically absurd.

If it's true that a rational human can move between different modes of thought, then we needn't think that the rules of science apply, without any change, to other human endeavours. The rules of the scientific mode don't apply (exactly) when I'm talking with my wife. And the rules of the scientific mode don't apply (exactly) when I'm reading the Bible, or praying, or thinking about Jesus' resurrection. When I'm thinking about the resurrection of Jesus, I'm not focused on questions about the mechanism of its occurrence. I'm asking different sorts of questions, and the rules of science are not directly applicable to answering those questions.

The upshot of the current conversation is: yes, science has rules, but no, accepting these rules does not preclude believing that miracles have occurred, nor that miracles could still occur. Recall, however, that we never argued for the claim that science has rules. Perhaps then we've misconstrued issues by basing the discussion on a false assumption. Perhaps science really just has one rule: find the truth.

Is it really plausible that science is guided by the single principle of finding truth? What then would we say about the many people throughout history who have tried to find truth in ways that have nothing at all to do with science? For example, many ordinary people think that there's as much truth to be found in a hike in the mountains as there is to be found in a textbook on condensed matter physics. It's just not clear that 'seek truth' is specific enough of a rule to guide scientific inquiry.

So, we maintain that the game of science does have rules. Sure, these rules can change, and these rules might not be universally accepted even at a single time. Nonetheless, we see no reason to think that accepting the rules of science mandates a global disbelief in miracles. Accepting these rules might preclude us from investigating the occurrence of miracles while doing science. But if we're aware that we've accepted such rules, then we're aware that we've constructed a net that might not catch all truths. In particular, this net might not be able to catch miracles.

There is another issue here that we must address: the possibility of *psychological tension* for those who attempt to do science, and also to contemplate the possibility of miracles. We're not talking here about any overt logical contradiction, because we are not talking any more about belief in particular propositions. Rather, we are talking about attitudes, stances, and lifestyle choices. The question is: Does a pro-science stance conflict with a pro-miracles stance?

Hans Halvorson recounts:

A few years ago, I saw a sign on a lamp post that advertised a 'learn to row' class. I remembered that my grandfather had rowed at college, and I thought to myself, 'perhaps this would be a pleasant way to maintain fitness.' So, I signed up for the class. Little did I know what I was getting myself into.

After a few short months, rowing had become like a religion for me. It was consuming much of my time, and no small amount of financial resources as well. It wouldn't be an exaggeration to say that my identity had changed: I was no longer just a husband, father, and professor. I was also now a rower.

In this age of specialization, it's not typical for serious scholars to have a serious non-scholarly hobby. Most professors aren't terribly interested in sports, and most sports aficionados aren't sitting up late at night to read philosophy. So, I sometimes feel like I'm trying to be two people at one time: during the early morning, and on weekends, I'm a rower. The rest of the time I'm a scholar. What's more, sometimes my two worlds collide.

For example, there's been at least one case where I turned down an invitation to speak at an academic conference because I wanted to row in a regatta. And even when there is no overt collision between these two lives, there's often a subtle tension in choices about how to spend my time, effort, and money.

Nonetheless, I find it wonderful to live simultaneously in two worlds. What's more, I've come to believe that I'm a better scholar because of my commitment to rowing. (As any rower will tell you, the sport certainly teaches life lessons.) Thus, I've come to welcome the superficial tension between these two lives that I'm living simultaneously.

I feel the same way about what it means to be a professionally scientific Christian. Certainly, there is some tension between the two lives. For example, on Sunday morning I might recite the Apostles' Creed:

> I believe in Jesus Christ, God's only Son, our Lord,
> who was conceived by the Holy Spirit,
> born of the Virgin Mary,
> suffered under Pontius Pilate,
> was crucified, died, and was buried;
> he descended to the dead.
> On the third day he rose again …

and my brain might say, 'But how did he rise again?' Or it might say, 'Am I sure that this really happened?' After all, extensive scientific training teaches one to check and check again before you're prepared to endorse a claim. This habit of mind, cultivated throughout a scientist's professional life, is bound to seep into one's religious practice. But note well: the same is true of one's interactions with other persons. If you spend a lot of time engaged in scientific investigations, then it's likely to change the ways that you interact with other people. In fact, many scientists will say that scientific habits of mind can be detrimental—if not properly regulated—for one's personal relationships. The same can be true of many other trained mental habits, including financial management.[6] Thus, the tensions of a rich and diverse life are present even for those who are not overtly religious.

The upshot of this discussion is that there is admittedly some pragmatic tension between the scientific mode of thought and the mode of thought that contemplates divine action, and in particular miracles. But as far as we can tell, the tension does not rise to the level of overt inconsistency. Even the most dedicated scientist has time to live a human life—and these human activities can make one a better scientist.

[6] See J. Welby, *Dethroning Mammon: Making Money Serve Grace* (Bloomsbury, 2016).

As we shall explain in the next chapter, we are not advocating an intellectual separation of professional and personal life, still less an intellectual inconsistency, but we are recognizing that complementary modes of thinking and acting are appropriate for different human activities. This is not a bug but a feature!

17

Wisdom and miracles

If you listen to some people, you might think that followers of Jesus are the most gullible people in the world. But that's an odd thing to think, because followers of Jesus are committed to loving God with all their minds—and that means, in particular, using their minds as tools for bringing God's love to the world. We think, in fact, that Christian people should be the most reticent people in the world when it comes to concluding that events cannot be understood scientifically. For Christians have a vested interest in understanding nature, and in bringing its ends into line with the good for humankind and all in our care.

17.1 Can Christians believe in miracles?

We aren't the first people to say that there are religious or theological reasons for questioning the existence of miracles.[1] For example, the seventeenth-century philosopher Leibniz argued that since God is perfectly rational, he would have created the world without any gaps that need to be filled in with miracles. Even back in the fourth century, Saint Augustine suggested that God planned all of the miracles in advance, and in fact planted the seeds for their occurrence at the moment of creation. And returning to modern times, several nineteenth- and twentieth-century German theologians, such as David Strauss and Rudolf Bultmann, argued that miracles wouldn't be a fitting mode of operation for a being worthy of the name 'God'.

The common thread in the thought of these miracle-deniers is a sense that a miracle would reveal some sort of *defect*—either in God's creation plan, or in God's execution of that creation plan. Suppose, for example, that Jesus performs a miracle to restore sight to a blind person.

---

[1] The phrase 'believe in' in the section's title is sometimes used in a strong sense, the sense of 'I think this is central and important', the way one might believe in the importance of freedom of speech, for example. However, in the present context by 'believe in' we just mean the humbler sense of 'holding the opinion that there can be such things'.

Well, says the miracle-denier, why did the person have to be blind in the first place? Is God patching up some defect in His creation? Wouldn't God in His perfect wisdom have arranged things in the right way from the beginning?

There are parallels here with Leibniz's criticism of Newton, which we discussed in Chapter 10. There Leibniz accuses Newton of diminishing God's greatness by supposing that God needed to intervene in the clockwork universe in order to adjust the orbits of the planets. For Leibniz, this kind of intervention would display a defect in God's initial creation.

A defender of miracles will be quick to point out that the reason for miracles is not necessarily to supplement some defect in creation. Yes, it may be that God could have created the world in such a way that the miracle wasn't called for. But God wasn't merely interested in the outcome of the miracle. God was interested in the process itself, in playing a more direct role in the events of the world at chosen moments. Other ends such as encouragement or teaching can thus be accomplished.

This kind of response leaves the defender of miracles open to a sort of moral objection. Would God really create people who will suffer (e.g. suffer blindness) just in order to swoop in and fix the problem—all in order to display God's glory? That picture sits uneasily with our moral intuitions. In reply to this one may argue that to create humans with weaknesses and pain, only to remedy some problems via miracles, is not worse than God simply creating humans with weakness and pain—which God has done (or at least, has allowed). Thus, the moral criticism of miracles doesn't settle issues.

There's another way to see this issue, a way that might seem to be suggested by recent advances in physics. Many people now believe that the physical world is genuinely indeterministic—i.e. that the future is not determined by the past. The thought here is that quantum physics describes a world in which events occur spontaneously, according to a statistical law, rather than according to strict deterministic law. Although some people—such as Leibniz, and also Einstein—would object to God creating a world with an 'open future', this consequence is welcomed by so-called 'open theists'.

Anyone who has studied quantum mechanics knows that it's a better theory for many purposes than classical physics. So couldn't God choose to create a world that obeys the laws of quantum physics—and hence is indeterministic? In that case, the perfection of God doesn't require God to determine all future occurrences at the moment of creation.

What's more, if God has left the future open, then there is one sense in which God can be 'surprised' by events. Imagine, for example, that God set up the perfect coin toss—where there was a 50% chance of heads, and a 50% chance of tails. Then God himself couldn't tell you which outcome would occur—provided that He doesn't intervene into the events. Nonetheless, in such cases, God would maintain the prerogative to intervene and guarantee a certain outcome. For example, if God wanted to coin toss to come up heads, then He could intervene to guarantee that it is the case.

Now suppose that a man developed terminal colon cancer as a result of a random genetic mutation. Then since (in this scenario) God created the world to contain randomness, He allowed there to be a chance that this man would develop cancer. But then, if God sees that the mutation has occurred, and that the man has cancer, then it would be perfectly reasonable for God to perform a miracle to heal him. The difficulty then becomes how we should understand the fact that God either universally or almost universally does not do that, but instead works with our willingness to learn science and medical practice and fair distribution of resources and help each other that way.

You may have heard Einstein's saying: 'God doesn't throw dice.' In response, the quantum physicist Niels Bohr is supposed to have replied, 'Don't tell God what to do!'[2] On this issue, we side with Bohr. We don't presume to know what kind of world God would create. We can say that God would create an 'orderly' world, but that's not a very specific claim. The best we can do is to try to make an educated guess about what the world is like—and then to go out and test our hypotheses.

But we don't even have to look to quantum physics in order to understand why a theist—and a Christian in particular—might think that God would perform miracles. For, other than some strict Calvinists, Christians believe that humans have free will. And if we have free will, then God might not be able to predict the course that the future will take. In this case, God might perform miracles precisely to redress some bad situation that has resulted from human choices. This is precisely how Christians understand the biggest miracle of all: the incarnation of

[2] Niels Bohr, 'Discussion with Einstein on Epistemological Problems in Atomic Physics', in *Albert Einstein, Philosopher–Scientist*, ed. Paul Arthur Shilpp (Harper, 1949), 211. Werner Heisenberg, *Encounters with Einstein* (Princeton University Press, 1983), 117.

Jesus Christ, his subsequent death, and resurrection.[3] Almost all Christians agree that human wrongdoing is not to be accounted for fatalistically. Rather, our wrongdoing results from our free choices. And that's why God needed to step in to redress the problem. God stepped into the course of history to be more deeply involved in His creation and overcome the problems that we created.

Of course, one could—as a movement in nineteenth- and twentieth-century liberal theology has done—think of the incarnation narrative as purely mythical, as a symbol for deep truths about humanity. But that would be to miss a couple of key points. First, it wouldn't approach the incarnation narrative as Jesus told us we should: like children. Second, it would mistakenly suppose that a mythological characterization of the incarnation narrative is the one that sits most comfortably with a scientific approach to the world.

The problem, we think, is that some philosophers and theologians understood science in an overly constrained way. They thought that the goal of science is to find the ultimate principles of things, the laws that govern all of reality. And they thought that even God had to conform himself to these ultimate laws. But we now understand science as working a bit differently than that. There's an aspect of science that the tightly constrained rationalistic picture misses: the fact that the order we find in nature is contingent. It could have been otherwise. In traditional theistic accounts, the patterns of nature were themselves instituted by God, and God is not himself constrained by those patterns.

We'll grant, however, that there is something right with these religiously motivated arguments against miracles. It has frequently been said that miracles are only useful for the spiritually immature—for those not yet attuned to God's voice. But once a person sees things from the right perspective, then the miracles are no longer necessary. This thought—that miracles are a concession to spiritual weakness—is supported by various things that Jesus himself said. For example, in rebuking the Galileans for their failure of faith, Jesus says, 'Unless you people see miraculous signs and wonders, you will never believe', and, in reply to requests for miraculous wonder-working on other occasions, 'A wicked and adulterous age asks for a sign'.[4]

---

[3] C. S. Lewis, *Miracles* (Collins/Fontana, 1947), Chapter 14.
[4] John 4:48; Matthew 12:39; 16:4.

Similar thoughts have been voiced by a number of other Christian thinkers, such as Augustine, Kierkegaard, and Barth. Augustine takes the point in a different direction, but a direction that is especially relevant for natural science. Augustine argues against identifying the miraculous (or wonderful) with those events that seem to run contrary to the lawful order of creation. On the contrary, we should see the lawful order of nature itself as a miraculous gift from God. 'Augustine's view is not so much that miracles are natural as that nature is miraculous.'[5]

Augustine provides a refreshing change in perspective from views that contrast miraculous events with the sorts of events that science can understand. Whatever other attractions such views might have, we're bothered by the way they split religious understanding apart from scientific understanding. The good thing about Augustine's view is that it imbues the scientific enterprise with a religious significance—when it might have been thought that engaging in science amounts to closing one's eyes to spiritually significant realities. Thus, Augustine's view serves as an important corrective even to the sort of view that we have voiced. We said that to see an event as miraculous is to see it as falling *outside* the domain of science. But that idea is, at best, a first approximation to an adequate understanding of the situation. It may well be that scientific practice calls for a narrow focus, and hence for a sort of exclusion of scientifically intractable features (e.g. spiritual significance of an event). But it's not as if our spiritual brains are left outside when we enter the laboratory. Similarly, it's not as if our scientific brains are left by the door when we enjoy the world from a sacramental point of view, or when we interpret events. At all times, these two perspectives are overlapping, interacting, and informing each other. The difference between them is a matter of emphasis, not a matter of a hard-and-fast distinction.

One of the failings that Christians or others may fall into at one time or another is to jump to conclusions about the lessons we find in the Gospel narratives. For example, there is a memorable account in Mark Chapter 2 and Luke Chapter 5 of a man lowered through the roof of a house in order that he can get close to Jesus in a crowd and be healed by him. Reading this, people have often taken as application the idea that we should see ourselves as like the friends of that man, so that when a friend of ours is in trouble such as paralysis or worse, our role is to 'bring them to Jesus' by prayer. But, in view of the rest of Jesus' teaching and his attitude to his followers, one might instead argue for another lesson here. Perhaps what Jesus has passed on to the community of his followers is not the role to be like those bystanders, but rather the role to be like Jesus himself, and actually heal people as best we can. On this view, the staff in the local hospital, and the many humble researchers who learned about the causes and cures of paralysis, have more fully taken on the mantle of being his followers, whether consciously or not, than have those whose only instinct is to pray and ask for miracles.

17.2 Straight dealing

When one looks into accounts of miraculous healing in modern times, wherever there is carefully gathered medical evidence it has been found to be consistent with the healing having taken place through natural, i.e. non-miraculous, processes.[6] Healing from a difficult and worrying condition will always be welcome, of course, and may carry special meaning for the person or people concerned. It may be providential. But this is not the same thing as miraculous in the sense that the healings reported in the Gospels are miraculous. Often the medical evidence is consistent with the sort of remission that cancer sometimes shows, or with an illness such as myalgic encephalomyelitis which is little understood and may remit for reasons that need not be miraculous. In view of this, one should be very cautious about what claims or promises are made to vulnerable people who may be hoping for help for themselves or a loved one.

Christian faith has a good correlation with positive health outcomes within the range of what may be called natural. The supportive and hopeful attitudes that a good Christian community fosters tend to have positive effects on both mental and physical health. The words in the letter of James, 'the prayer of faith will save the one who is sick, and the Lord will raise him up' are helpfully open. They express quite well the sort of encouragement and sense of being cared about that people get when others pray for them, but they do not guarantee healing. They only suggest some sort of general positive effect. James was, we

---

[6] Peter May, 'Miracles in Medicine', *Science and Christian Belief* **29** (2017), 121–34.

think, being careful to avoid overstating the situation. He wanted to offer some helpful guidance without being misleading or untruthful.

The miracles described in the Bible take place mostly in groups, the main groups being associated with the liberation from Egypt, with Elijah, and of course with the life of Jesus. It is not possible to count them unambiguously, because the count will depend on questions of interpretation, such as whether the story of Jonah is historical or parabolic. Also, sometimes an event might take place owing to natural causes such as an earthquake, but with a timing that proves significant for the people concerned. The crossing of the Jordan may be of this kind (see Chapter 19). However, even if one includes events of significant timing under the title 'miracles', then the total number of miracles described in the Old Testament is some tens or perhaps around one hundred, and they take place in groups. It follows that in a typical year in the life of ancient Israel, almost certainly no miracles at all took place. We are dealing with a period of well over a thousand years, and most of the marvellous events take place in a few of those years. Even allowing that the record is far from complete, miracles in ancient Israel were not just rare but extremely rare. Most people could expect not to encounter one or even hear of one happening elsewhere in their whole lifetime.

In the apostolic period described in Acts of the Apostles, there are beautiful and remarkable events, but nevertheless, most people experienced the amount of illness or calamity and death that is normal in human life. Paul had some sort of difficult physical ailment, and others had equally serious illness and setbacks. All died.

If we designate as 'miraculous' events that don't fall into the normal patterns of the world—the patterns that science elucidates—then the miracles that actually happened (as opposed to mistaken reports or misunderstandings of natural phenomena) are pointers to or signs of what the larger patterns of reality actually are. Such events are not a breakdown of a given law, but a breaking-in of a higher law. This is like the use of dissonance in the hands of an expert composer, or the use of a deliberate missed beat in the flow of a poem. Without the established normal flow the exception would not be possible, but when introduced at the right moment, such a disruption is where a larger meaning may be at work.

The role of miracles in the life of Christ is centred on their significance in indicating the unique importance and role of the person at the heart of them. The beautiful events that took place were telling pointers, but they did not address general healthcare. The rate of cancer, leprosy, haemophilia, blindness, and so on in the general population did not change. To address those issues, the crucial factors have been hygiene, fairer distribution of wealth, scientific understanding of causes and cures, and simply the recognition that every individual is valuable. These are the things that Jesus made possible, in the long term, by his teaching and demonstration of what our general attitudes should be. Through modern-day infrastructure in developed countries, women can have every reason to hope for safe child delivery, and many illnesses which were widely experienced only a couple of generations ago (tuberculosis, diphtheria, polio) are all but eradicated. Even some forms of blindness can be healed by laser eye surgery. Human potential is truly marvellous when we live out the sort of partnership with God which Jesus made possible.

What does partnership with God look like in practice? It does not take the form of individual people getting super-powers. It takes the form of communities of people coming together with greater levels of wisdom, generosity, trust, and mutual forgiveness. This in turn leads to the decision to offer education (including literacy and numeracy) as a universal freely available start for all, and it should lead to empowerment of women equally with men, and to culture with good governance, which makes possible the long-term stability that liberates some to study science as well as other creative endeavours. The theme of partnership motivates passion for truth and intellectual integrity, and an interest in the wider world and how it functions.

Jesus spoke strongly against the craving for miraculous signs and wonders, which appears to have been a common fault in first-century Israel, and it can be equally common in some settings today. It seems to involve a combination of arrogance on the one hand and faithlessness on the other. Arrogance, because it tries to get God to play tricks before someone will commit to learning from Him. Faithlessness, because it is a desperate search for reassurance. But when people courageously commit themselves to loving others in God's name, with or without signs and wonders, then time and again marvellous things happen.

17.3 Cessationism

Our view of miracles may become more clear if we discuss a theological issue. Among Christians, there is a divide between 'cessationists' and 'non-cessationists'. In short, cessationists believe that after the biblical canon was settled, there was no longer any reason for miracles to occur; i.e. the point of miracles was to ratify the revelation from God through the prophets, Jesus Christ, and the Apostles. But once this revelation was written down in the books that were to form the Bible, there was no more reason for God to interact with nature in this way. Therefore, say cessationists, miracles no longer occur.

Cessationists tend to be Protestant Christians. Catholic Christians cannot, strictly speaking, be cessationists, since the Catholic Church requires ongoing miracles in order to confirm that a person is a saint. But not all Protestants are cessationists. Many so-called charismatic or Pentecostalist Christians believe that God still performs miracles. Christians who believe in 'faith healing' are not cessationists.

Since this book isn't a contribution to academic theology, we will refrain from a deeper engagement with the arguments for and against cessationism. (We've seen good arguments on both sides of the question.) We'll just say that when we put on our scientists' hats, so to speak, we don't distinguish two epochs—an epoch (before the sealing of the canon) in which miracles occurred, and another epoch (up until the present day) in which miracles do not occur. But that doesn't mean that we interpret the biblical miracles through the lens of science. Indeed, the moment that we shift our attention to the spiritual significance of the biblical miracles, we have ceased trying to engage in precise, controlled, scientific inquiry. When we read the Scriptures, we are told that God was active, which means the following to us: the tools of natural science are no longer appropriate for a full understanding of these events—or, at least, for the kind of understanding that is being offered to us.

You might wonder, then, would we (the authors) be willing to describe events in our own lives as miraculous? It depends. At the beginning of the previous chapter, I (HH) described the events surrounding the birth of my son. What I didn't mention was that his birth itself was something of a miracle. You see, during the first trimester of my wife's pregnancy, she developed a large blood clot in her uterus. The blood clot was so large that it was putting pressure on the developing baby, and doctors said that he had less than a thirty percent chance of survival. They then prescribed bedrest to my wife, and warned her that any sort of exertion could cause a miscarriage.

The next seven months of bedrest were like a prison sentence for my wife. She felt as if she was waiting for her execution—only the death of a child may be worse than execution for a mother. Nonetheless, the days passed, and at each biweekly doctor's appointment, the heart beat remained strong. As we reached the eighth month, we could hardly believe it. Could we allow ourselves to hope?

As we celebrated our son's tenth birthday, my wife posted the following on Facebook: 'Our miracle baby is 10 years old!' How should we understand what she was saying? Can we paraphrase her statement as saying that she believes that the laws of nature were somehow violated during her pregnancy? I know my wife, and I can assure you that she wasn't suggesting this. Could we then paraphrase her statement as saying that, although the laws of nature were most definitely not violated, we are filled with a warm fuzzy feeling about our son? Once again, no: I think it would do violence to the meaning of her statement to say that it has a purely metaphorical meaning. No, she wants you to understand her statement in a way that has nothing to do with laws of nature, or with any other attempt to say something (either positive or negative) about the mechanisms involved in the baby's gestation. She was suggesting that she attributes our son's existence to the power of God, a God who is not Himself subject to the vicissitudes of created reality.

17.4 Rational belief in miracles

We now come to a piece of thinking that we (the authors) found hardest to put into words and discussed the longest, compared to any other item in this book. So if the following remains somewhat open or unfinished, that is why. We want to express our sense that to believe that certain events have been miraculous is a rational thing to do, but it is not a conclusion that follows inexorably from a line of reasoning, and it is not an item that can be surveyed and chosen, or not, dispassionately, as if we were disinterested. To admit that something was miraculous is a highly engaged and interested thing to do. In the cultural climate of modernity and post-modernity it can also be a courageous thing to do, and sometimes it may be an acutely humbling experience.

One of the central claims of Christian witness is the resurrection of Jesus. This is not merely the report of a man alive again after he had died, but rather the inauguration of a creative transformation in human history, and a previously unrealized mode of embodied existence. This claim requires of us careful intellectual study, but in the end this is not a case of intellectual compulsion, but rather a situation involving a decision of commitment. It is not like investigating other questions of either science or history. It is too personally significant, and requires of us not just intellectual assent or dissent, but the decision to embark on a new way of being human. No external argument can do all the work of carrying one to the place of Christian commitment. But that commitment remains fully rational and meets all the demands of rationality.

A follower of Jesus ought to apply the most exacting epistemic standards to himself. He must be wary of wishful thinking, and he should strive to bring his beliefs closer into alignment with reality. To this end, he should take advantage of all the evidence he can lay his hands on to test and amend his beliefs. It's not surprising then that many scholars have felt it important to investigate the question: did Jesus rise from the dead? Some scholars argue that the evidence points toward a negative conclusion, and other scholars claim that the evidence points toward a positive conclusion. We are grateful for this careful and detailed scholarship, which provides a resource for our epistemic duties as Christians. But the question of Jesus' resurrection differs from the sorts of questions that historians and scientists ask in their normal professional capacity. We would argue that investigations into Jesus' resurrection cannot meet the ethical and epistemic norms of scientific inquiry, because every investigator is too heavily invested in the outcome.

Imagine trying to answer the question 'Is my life in harmony with reality?' by means of a fully objective scientific investigation. It can't be done, first because it is not the kind of question which is amenable to scientific methods, and second because you have a vested interest in the outcome of that question. You already have a life in which you have made choices with consequences. As a result, you are not qualified to investigate that question with appropriate scientific objectivity. Indeed, nobody is qualified to investigate that question for him- or herself objectively. Everyone has a conflict of interest.

Now, if the question 'Is my life in harmony with reality?' cannot be investigated by anyone in a purely scientific manner, then neither can the question 'Did Jesus rise from the dead?' For an answer to the latter question bears directly on the former question—at least for people who understand the gravity of the resurrection claim. Everybody has huge stakes in the game. If the investigation yields the verdict that Jesus did *not* rise from the dead, then the Christian has based their life on a lie. If the investigation yields the other verdict, then it calls the investigator to start a new life as a follower of Jesus. In either case, the individuals involved have too much to gain or to lose. In ordinary scientific practice, such an extreme conflict of interest would disqualify people from participating in the investigation. In the case of Jesus' resurrection, everybody has such a conflict of interest.

But this does not mean one should not investigate at all! Far from it! Since the evidence stubbornly refuses to evaporate, one should indeed investigate with great care, and it would be a serious failure of intellect to disqualify the evidence on the basis of a prejudice about other possibilities, for example by making the assumption that it is bound to be more likely that all the human reports are mistaken than that something outside our normal experience occurred. We (the authors) each investigated in so far as we could. We concluded that the evidence is striking, but also that reason alone does not close the matter down in either direction. We are left with the right and the responsibility of making a choice, of opening ourselves to one life or another.

The respected New Testament scholar N. T. Wright has undertaken painstaking historical analysis which we find to be immensely valuable.[7] When he asserts, in his introduction, 'the *only* possible reason why early Christianity began and took the shape it did is that the tomb really was empty and that people really did meet Jesus, alive again', we take this as hyperbolic for emphasis. In context it was perhaps intended as a shock tactic to wake up those who assert that no sane person believes in the resurrection, to help them to open their mind to the possibility that God the Creator has launched his new creation in Jesus. Wright thus alerts the reader to the fact that the situation is not as self-evident as contemporary culture often assumes. And equally, neither is Christian commitment a self-evident conclusion that some sequence of argument can draw for us. Historians and anthropologists have found many different reasons for the origin of the Christian religion—socioeconomic and psychological reasons, among others. But the insights available from those explanations can be taken on board, and we find that they do not undercut our conviction that Jesus did rise from the dead. We don't believe that Jesus rose from the dead merely because we were unable to find other ways to account for the data. Belief in God requires choice and commitment. We made the choice because we were allowed to in integrity, and felt called upon to walk this way.

---

[7] N. T. Wright, *The Resurrection of the Son of God* (Fortress Press, 2003).

17.5 Miracles and evidence

Now let us apply some of the philosophical points we have been making to the wider discussion of miracles. We have pointed out that this is not nor can it be an essentially scientific issue. Why do some people nonetheless insist that there is conclusive scientific or historical evidence that miracles occurred? We suspect that their insistence is based on a worry that if there is *not* scientific evidence that something happened, then believing that it did happen is irrational and perhaps wishful thinking. And nobody wants to admit to being irrational—hence, the insistence that the evidence demands belief in miracles.

While this attitude is understandable, we suspect that defenders of 'scientific evidence for miracles' have not really thought carefully about what they are saying. What exactly would it mean to find scientific evidence that a miracle occurred? Typically, science looks at particular events to either confirm or disconfirm general hypotheses. So, for example, seeing a black raven would count as evidence for the hypothesis 'all ravens are black'. Similarly, seeing a white raven would disprove the hypothesis 'all ravens are black'.

The claim 'a miracle occurred' is not a generalization of this kind, but it would be easy to confirm it: we only need to witness one event that was miraculous. But an event's being miraculous is not like a liquid's being acidic: we don't have litmus paper for the miraculous. So, what are the features of an event that give the scientific proof that it is miraculous? I'm sure that sophisticated defenders of 'scientific proof for miracles' have an answer to that question. But since they're not here, we will have to try to answer for them. We guess that they would say that the tell-tale sign of the miraculous is a violation of the laws of nature. In other words, the reason to believe that an event is miraculous is because that event violates the laws of nature.

Setting aside our earlier worries about the laws of nature, this proposal is still a bad idea. In good scientific practice, if we witness a violation of the laws of nature, we don't conclude that a miracle occurred—we conclude that we were wrong about the laws of nature. Such anomalous events provide us with crucial clues about how to construct better theories. Think, for example, of the experiments in the late nineteenth century which ran contrary to the laws of classical physics. The great scientists of the time did not conclude that they had witnessed miracles. Instead, they concluded that classical physics is false—and their successors went on to create a better theory, which we now know as quantum physics. As we were reminded at the end of Chapter 12, when we encounter something which we do not understand in science, the correct response is not to invoke God but to become better scientists. We are best placed to respond in worship to those aspects of the creation which we do understand, not to those which we don't.

There's another problem with a simple conception of scientific evidence for miracles. The problem is that when it comes to spiritually significant events, we all are prone to a certain sort of blindness. As Jesus himself said,

> This is why I speak to them in parables.
> Though seeing, they do not see;
> though hearing, they do not hear or understand.[8]

In the context, he seems to be saying that he is doing his best to convey a message to people who are thinking in mistaken ways that make it hard for them to grasp the message. In the spirit of what Jesus said, our own psychological resistance can prevent us from seeing miracles for what they really are. It is only by God's grace that our spiritual eyes are opened.[9] To be present at the occasion of a miracle, or to make oneself present at one step removed through reading or hearing of it from a trustworthy source, is comparable to being present at the moment when someone opens up their heart to say, 'I love you.' This is not the moment to respond by saying, 'How interesting: look, your neurons are pulsing and your tongue is flapping.' But if we refuse to accept that something more personal and intimate is taking place, then that is liable to be the sort of response we will make.

Let's not take that too far—there is a point in saying that there is scientific evidence for miracles. That point is that these events actually did occur, and could have been witnessed. For a better understanding of how these events relate to scientific theories, it might be helpful to consider the notion of a 'scientific revolution' as described by the historian of science Thomas Kuhn.[10] According to Kuhn, there are moments when the rules of normal science break down—when science as we know

---

[8] Matthew 13:13.

[9] Our view—that observing a miracle requires faith—is shared by the Catholic philosopher Ralph McInerny. See his *Miracles: A Catholic View* (Our Sunday Visitor, 1986).

[10] T. S. Kuhn, *The Structure of Scientific Revolutions* (University of Chicago Press, 1962).

it just won't help us understand an event or events. If fortune favours us, then some day we might be able to build a better scientific theory that makes sense of these events. But as things currently stand, they simply don't fit into our scientific picture of the world.

Miracles are, in one sense, like these sorts of 'science-breaking events' that Kuhn describes. When a miracle occurs, it doesn't confirm one scientific theory (the theism theory) over another theory (the atheism theory). No, a miracle is the kind of event that refuses to be crammed inside any human-built theory. It's the kind of event that demands a response different in kind from the typical scientific response of attempting to classify and analyse.

You might say: 'But the sorts of event that Kuhn describes are all perfectly natural events.' Ask yourself: what's the source of your intuition that these events are natural? The reason we think those events are natural is because a *subsequent* scientific theory gave us a way of understanding them. For example, consider the sort of anomalous event encountered by Faraday, when he witnessed iron fillings spontaneously aligning themselves. Was that event natural? Well, we think so now— but that's because we understand it in terms of the electromagnetic field (a concept which Faraday introduced to account for his observations). Many events that we now think of as ordinary would once have been considered to be quite miraculous.

Don't mistake our intentions here. We are not saying that the biblical miracles are completely natural events that ought to be explained by some future scientific paradigm. As we shall see in Chapter 19, natural causes can be suggested for some of the biblical miracle stories.[11] But their significance for us does not depend on finding such causes. Our belief that God acted in history is not conditioned on our having some quasi-scientific story about how He did it, and we don't rule out that God could have done such things even if they completely defied the laws of nature. Our purpose in bringing in Kuhn's observation about scientific progress is to point out what learning can be like. The case of notable reproducible phenomena illustrated by Faraday's experiment, and the case of events properly called miraculous, have in common that both offer invitations to learn more about the nature of reality, and both will ultimately take their place in a fuller picture. In that fuller

[11] C. J. Humphreys, *The Miracles of Exodus: A Scientist's Discovery of the Extraordinary Natural Causes of the Biblical Stories* (HarperCollins, 2004).

picture, the distinction between 'natural' and 'supernatural' is probably not the right one to try to draw. The question of miracles is not a question of laws or breaking of laws, but a question of wisdom in response to what we each experience and what other people report to us of their experience.

18

You can't live a divided life

Hans Halvorson gives a personal account

I grew up in two cultures—the first of which worshipped science, and the second of which denigrated it. It's really not surprising that I'd end up working in philosophy of science, if only to resolve the cognitive dissonance.

The United States is deeply divided, in so many ways. We worship science, not so much for the knowledge in provides, but for the lifestyle advantages. We also cling to the old religious identities that we inherited from our ancestors in Europe, Africa, and Asia. We've never been very good, as a culture, at bringing these two sides of ourselves together.

But no human can continue to live a divided life. One side or the other will give way. In my childhood community—an offshoot of American fundamentalist Christianity—it was the scientific side that ultimately gave way. My community loved what science does for us. But when it came to the ultimate nature of reality, we turned to the Bible. (I have come to realize, in fact, that many people in the Christian community in the USA are scientific anti-realists. Taking this position allows them to use science without worrying that it will conflict with their religious beliefs.)

The community in which I was raised was by no means anti-intellectual. In fact, compared to many of my secular friends, it seems that the community in which I was raised was extremely cerebral. My parents' house was filled with books, by all sorts of authors—both religious and secular. We were encouraged to read widely. We were taught that logic is a gift from God, and that the truth will always have the better argument. Discussion over the dinner table often turned to philosophical questions, such as free will or mind–body dualism.

It was a good time to grow up in the USA. The religious roots were deep and strong, but material bounty and educational attainments had reached new heights. My mother was a science teacher with multiple graduate degrees. My father—an investment banker by trade—took night classes in theology, and amassed an extensive library of classical works in philosophy and theology.

I'm deeply grateful for the intellectual heritage that I received. However, the empirical sciences remained a sort of second class intellectual occupation. Science dealt with ephemeral truths. For eternal truths, one had to turn to God's revelation to humankind in the Christian Bible.

As a teenager, I realized that I had a proclivity toward selfishness, and for pursuing ephemeral pleasures over the greater goods of a human life. It's a proclivity that all humans have, some to a greater extent than others. I came to conclude that I was like a sick person, who needed a doctor. I found that my 'dear self'—as Immanuel Kant called it—just didn't want to be demoted from the little throne that he had built.

What I needed was a radical reorientation. I wouldn't have used those words at the time—they only come to me after much reading of philosophical accounts of conversion. But they describe the situation accurately.

I started praying that God would change me. And God answered my prayer, decisively and radically. If you had asked me, when I was fourteen years old, what I hoped for, I probably would have answered: 'fame, money, entertainment, and an easy life'. If you had asked me the same question when I was fifteen years old, the answer would have been very different indeed.

I'm sure that psychologists could provide an interesting naturalistic story about what happened to me when I was a teenager. And their story might be true, for all that I know. Perhaps I needed to find a coping mechanism. Perhaps humans have evolved an altruistic consciousness, and it was asserting itself through me.

What difference should it make to me that this spiritual change might admit of some sort of scientific explanation? Would I be forced to say that my sense of moral inadequacy was illusory? Would a scientific explanation reduce the personal significance of these events?

Honestly, I don't really care what mechanisms, psychological or biological, were implicated in my personal reorientation. For me, from the first-person perspective, the ability to change was the most precious gift that I have ever received. It originated from outside of me. I didn't simply try harder, or make incremental moral improvements. The reorientation was radical—and it was noticed as such by the people living around me. To echo the words of a famous hymn, I 'was blind but now I see'.

Being an unusually serious teenager, I didn't want to spend my life chasing ephemeral ideas. I wanted to find those truths that would never change. I enrolled in a Christian college, thinking that I would probably become a theologian. But there were signs that I needed a change of plan.

The first sign was that I had a sort of recalcitrant talent in mathematics. I say 'recalcitrant', because I was scarcely interested in using this talent. It's just that I was good at mathematics—my teachers said I was a natural.

The second sign was that I found the academic study of theology didn't permit me to explore the full range of my interests. I was interested in literature, languages, and aesthetics. And I was soon to discover an extremely strong interest in mathematical logic.

The third sign came in the form of a personal relationship. In the first philosophy class I took, there was a girl who sat on the opposite side of the room. I noticed on the second day of class when she gave a particularly clever answer to one of the professor's questions. We began a friendship that has lasted until this day—and that turned into love, marriage, and children.

But what does love have to do with theology? In my case, love made me question the theological framework that I grew up with—a framework that included a prohibition against women teaching in the church. I became convinced of the incorrectness of that view, and accordingly, I moved away from thoughts of pursuing theology—at least in the tradition in which I was raised.

There were two other points of light in the form of two philosophy papers that I read in my second year of college. The first was a paper by Alvin Plantinga on 'reformed epistemology'. The second was the famous 'Two dogmas' paper by the philosopher W. v. O. Quine. Reading these papers was a sort of conversion experience for me. I realized two things. First, Plantinga showed me that Christians could do groundbreaking intellectual work in the secular academy. Second, Quine showed me that philosophy could be clear, precise, and integrated tightly with science and mathematics. Within a span of a few weeks, I was converted to the idea of a career in philosophy.

Of course, my intellectual road since then has been long, with many unexpected twists and turns. I could not have imagined, for example, that physics would come to play such a central role in my intellectual life. I had loved physics when I was in high school, and I had especially enjoyed thinking about foundational issues. But that seemed, at the time, like a sort of game, without any lasting significance.

It was in the philosophy PhD program at the University of Pittsburgh where I saw the most beautiful model of science–philosophy integration. In the mid-1990s, Pittsburgh was like philosophy of science heaven. John Earman and John Norton were at the height of their powers. The philosophy department then hired Rob Clifton and Laura Ruetsche, who also worked in philosophy of physics. On top of all that, the physics department at Pitt had a young relativist named Carlo Rovelli, who was as comfortable talking about philosophy as he was about loop quantum gravity.

The icing on the cake for me was when my father sent me two books. The first was a book by Dr John Polkinghorne, the physicist turned Anglican priest. The second was a book by Professor Thomas Torrance, the Scottish theologian who also had a PhD in physics, and who had written some interesting things about the history of physics.[1] I found it beautiful how these two men brought together what I had seen as separate worlds—the world of theology and the world of physics. Instead of seeing physics as merely a tool to build bridges and smart phones, they saw it as a way of orienting oneself properly toward creation. Doing physics wasn't merely a means to an end, it was a genuine source of joy, and even of illumination.

For me, physics was beautiful, but I hadn't been able to give my heart to it. My heart was devoted to 'ultimate questions'. But Polkinghorne and Torrance broke down the barrier between the ultimate questions of theology and the penultimate questions of physics. Having read those two books, I was free to pursue science.

Of course, it was too late for me to become a real scientist. In this age of specialization, to rise to the top of a scientific field, you've got to be extremely focused from an early age.

Each of us is placed in a unique situation, and often that situation appears to be less than ideal. My situation didn't seem too great. Basically, I woke up one day and realized that I should have chosen a career in science, and physics in particular. I loved physics, and Polkinghorne and Torrance showed me that the scientific calling was no less noble than a calling to do theology or philosophy. (In fact, Polkinghorne and Torrance

---

[1] I subsequently bought several more books by Polkinghorne and Torrance, with the result that I cannot remember which books were the gift of my father. All of them are heartily recommended.

would say that's a false dichotomy. The scientist is called to do theology, and the theologian is called to do science as well.)

Nonetheless, my parents had taught me that 'God has a reason for everything.' Thus, while there were certainly days when I wished that I was a physicist, I thought to myself: 'Surely I can make the best of this weird situation. Here I am, somebody with a decent education in mathematics, a bit of physics, and a lot of philosophy. I love the big questions that come up in philosophy, but I prefer the rigorous techniques of the scientists.'

The answer was right before my eyes: I would become a philosopher of science, and more specifically a philosopher of physics.

I didn't fully realize at that time just how specialized I would become in philosophy of physics. By the time I was finishing my PhD dissertation, I had become something of an expert on algebraic quantum field theory. (The topic was in the air in Pittsburgh, and my mathematical background made me a good candidate to work on it.) I was proving theorems about Bell's inequality, and the only time I thought about theological issues was on Sunday mornings (if I could prevent my mind from wandering during the sermon).

As any scientist will tell you, scientific projects take on a life of their own. Once you solve one problem, five new problems emerge. In my case, some early successes with algebraic quantum field theory led to my spending about a decade thinking about little else. My Christian faith remained, although I grew increasingly humble about the ability of my mind to reach deep insights about God, and his reasons for creating the universe. Science taught me humility. We know so little about the universe. If we know something about God, then God surely must have done something extraordinary to make that possible.

I longed for a chance to talk to some smart people about issues of faith and science. But there just weren't the opportunities for that within my normal academic life. Fortunately for me, word leaked that I was a Christian, and I started to get some invitations to write and speak about issues in faith and science.

If I recall correctly, it was in the summer of 2008 that I first met Andrew Briggs, when we both attended a meeting of the Board of Advisors of the John Templeton Foundation. But both of us were overwhelmingly busy at the time—so after a brief introduction, we fell out of contact for a few years.

Andrew Briggs and I met again when I visited Oxford in 2011. At that time, we started talking about ways of connecting science, philosophy, and theology. Anyone who knows Andrew B. can tell you that he is a visionary and also an indefatigable organizer. He loves to match interest with opportunity, and he hit the right note when he asked me if I would like to join the 'Experimental Tests of Quantum Reality' project. He hit the right note again when he said, 'I think you would enjoy talking with my colleague Andrew Steane.' This book is the result of the very enjoyable conversations that began that week in Oxford, and of all the thinking that we each individually had done on these topics over the years before.

19

Learning from the Bible

Several of the issues that have arisen in the course of this book touch on an aspect of Christian discipleship which is easily misconstrued and is hard to get right. This is the issue of how we should treat the Bible and receive what it offers. We will not attempt a thorough discussion of this, but it seemed to us (the authors) that it is a subject that comes up in some of the chapters of the book, and therefore some general comments would be appropriate.

The main thing we have to say is that we don't know the answers to all the questions that come up in handling the Bible, but we want to affirm that *it is alright not to know*, and that living with this kind of uncertainty need not prevent one from learning.

The uncertainty concerns partly the limited knowledge of the natural world that the texts in the Bible display, but more importantly it concerns how to handle the more violent parts of the Bible, and the parts where morally questionable attitudes or actions are attributed to God.

19.1 The Bible and physical facts

Readers have sometimes found it unsettling that the Bible has been written by people who had no way to know, and did not know, various facts about the natural world, such as the very great age of the Earth, the vastness of the universe, the causes of disease, and the fact that humans have a common ancestry with other animals. However, one does not need to know about those things in order to learn the priority of forgiveness over revenge, and of justice over bribery, and the equality before God of people of all kinds, the importance of humility, the value of honest enquiry, and so on. Therefore, the limited scientific knowledge on show in the Bible need not be a problem. Our knowledge of the cosmos is still limited now, with great uncertainty over the composition of most of it (the dark matter problem), and deep puzzles over the nature of physical reality (the quantum interpretation problem) and many other open questions, but all this is mostly irrelevant to the task of configuring our relationship to God and to one another correctly. The important point is that the central argument of the biblical texts is not based on, and does not rely on, a wrong statement of physical facts.

The way people have interpreted the Bible has often resulted in structures of doctrine which are based on wrong statements about physical facts. Such structures must be abandoned when the thing underpinning them has been shown to be false. To make the point, we can use a simple example. If anyone's doctrine is built on, or requires, the claim that the Earth is less than a billion years old, then one should re-examine the basis of that doctrine, keeping in mind the physical evidence. One does not first study Scripture, and then bring in science. Rather, one studies Scripture in such a way that scientific knowledge is one of the hermeneutical tools (*hermeneutics* is the branch of knowledge that deals with interpretation, especially of the Bible or literary texts), alongside other tools. That does not mean that we interpret an ancient document to say things that were unknown to its authors; rather, it helps us to understand what aspect of genuine experience and knowledge the authors were writing about.

In the example of the age of the Earth, the outcome is straightforward. The opening chapter of Genesis uses the metaphor of 'days' as a literary device to talk about organized, measured activity, and the great age of the Earth and the wider universe is a welcome further ingredient to the picture. By keeping the science in mind one gets more confidence about the hermeneutic, and one is less likely to make mistakes about what can be appropriated from any given passage of Scripture.

The box gives examples of writers who, long before the modern scientific age, recognized the subtlety of descriptions of creation, and how we need to look beyond the Bible for what we would now call scientific accounts. These writers were careful not to build structures of doctrine on incorrect statements about physical facts. Another example appears in Psalms 96 and 104, which both state that the Earth 'stands firm and shall never be moved'. What are these psalms asserting through their poetic statement? Is the psalmist celebrating the fact that the basic support of our physical existence is not about to break or dissolve the way human empires do, or is the psalmist saying that there is no sense in which the Earth as a planet could be in motion as a whole, relative to some other reasonably natural choice of reference frame? It is only the advent of scientific study that made the second thought even possible

Some examples of pre-modern writers concerned with the interaction of natural philosophy (such as astronomy) and biblical texts

- Augustine of Hippo (354–430) in his commentary on Genesis (*de Genesi ad litteram*) develops an idea of two moments of creation, one primordial or 'outside time', the other within the course of development over time. He wrote, for example, 'One will ask how they were created originally on the sixth day. I shall reply: "Invisibly, potentially, in their causes, as things that will be in the future are made, yet not made in actuality now."'[1]
- John Philoponus (c.490–c.570) in *De Opificio Mundi Libri*. Philoponus introduces his commentary by trying to establish what kind of questions Genesis is and is not addressing. His conclusion is that the purpose of Genesis is not to provide a scientific account of the origin of the universe, but 'to teach the knowledge of God to benighted Egyptians superstitiously worshipping the sun, moon and stars', and this is a theme he returns to several times in the course of the book.[2]
- William of Conches (1085–1154): 'The authors of Truth are silent on matters of natural philosophy, not because these matters are against the faith, but because they have little to do with the upholding of such faith, which is what those authors were concerned with.'[3]
- Nicole Oresme (1325–1382): although he subsequently decided that Ps 93:1 was evidence that the Earth does not rotate, he also took passages figuratively 'by saying that this passage conforms to the normal use of popular speech just as it does in other places...which are not to be taken literally.'[4]
- John Calvin (1509–1564), the Protestant reformer and biblical scholar, wrote a commentary on Genesis approximately seventy years before Galileo's conflict with the Roman Catholic Church when the 'Bible and science' debate began to get heated. He wrote, 'He who would learn astronomy and other recondite arts, let him go elsewhere.'[5] Calvin bases this on the nature of the language in Genesis 1 and its purpose.

---

[1] Augustine of Hippo, *de Genesi ad litteram*, 6.6.10.

[2] Roger Wagner and Andrew Briggs, *The Penultimate Curiosity: How Science Swims in the Slipstream of Ultimate Questions* (Oxford University Press, 2016), 121.

[3] James Hannam, *God's Philosophers* (Icon Books, 2009), 62–3.

[4] Hannam, *God's Philosophers*, 187.

[5] J. Calvin, *Commentary on Genesis* (1554), on Gen. 1:8.

to be contemplated. It is not a thought that could ever have entered the psalmist's mind. Therefore, the second version cannot be what the psalm means.

We mention this rather easy example because it is the very one which people found difficult when Copernicus and others first proposed that the Earth is in translational and rotational motion relative to the Sun. Some leading Christian thinkers saw quickly that there is no contradiction with the Psalms, but others thought that there was. The error they fell into was to project back onto an ancient author a way of thinking that that author did not have and could not have had.

This leads us on to our next point.

In Section 5.5 we invited the reader to imagine how one might describe the creation poetically to an Ancient Near Eastern mindset. Genesis Chapter 1 was written for such a Hebrew community. It is often assumed that they would have taken it literally, as if this was an ordinary week of twenty-four-hour days, and now in our time we are being told to take it another way. But there is no need to assume that the original community thought in terms of a literal week when they wrote or listened to the account. They probably did not think that way. The very way people think, their patterns and assumptions of thought, are profoundly shaped by underpinning culture. The power of science, working away in the background to people's world view, began to suggest that thoroughly literal readings of Scripture were somehow more honourable or more powerful or more true, even when the literature does not have the appearance of prose or straightforward history.

A similar mistake can be made when people assume that biblical authors had a 'three-tier cosmology' in the sense of heaven above, Earth below, and *sheol* or *the grave* below that. But this is to impose either a medieval European way of thinking or a modern Western way of thinking (i.e. getting everything pinned down in terms of physical location) onto a Hebrew community which did not think that way. The biblical documents have a fluid view in which sometimes God is located throughout the natural world, sometimes in a tent, and then immediately not able to be contained in any structure, and then 'hearing from heaven'. In short, they are reporting human experience, and they simply could not care less about the subsequent 'three-tier cosmology'.

On our bookshelves, we (the authors) have many different genres of books: poetry, biography, history, compilations of letters, philosophy, imaginative writing, and mathematical science. The different books in the Bible contain examples of most of the genres, and others besides, except not mathematical science. The necessary intellectual equipment was not available to the writers, and no one at the time would have understood it. Even today the majority of the population would not be able follow advanced texts of quantum theory, general relativity, plate tectonics, or systems biology. Each part of the Bible must be read with respect for the genre of literature that it represents and the intention of its author in writing it.

19.2 Remarkable events

The biblical texts report many remarkable events as well as more every-day ones. Some of these events are commonly called miracles; others are not commonly called miracles but are remarkable nonetheless. An example of the first case is the crossing of the Jordan; an example of the second case is the deliverance of Jerusalem from an Assyrian siege.

The crossing of the Jordan River was indeed remarkable. When in flood, the river could be fast-flowing and up to a mile across, a truly impenetrable barrier to any large group of people. Yet the record describes the following:

> As soon as the priests who carried the Ark reached the Jordan and their feet touched the water's edge, the water from upstream stopped flowing. It piled up in a heap a great distance away, at a town called Adam in the vicinity of Zarethan while the water flow-ing down to the Sea of the Arabah (the Dead Sea) was completely cut off. So the people crossed over opposite Jericho.[6]

This account was received at face value by most readers of the text until the modern era. However, in modern times we have become aware of the fact that ancient literature can contain legends. For a river to stop flowing and then start again after people had crossed over is the sort of thing that one would ordinarily suspect is a legend. So there arose a division in the scholarly community. Most came to the view that this was a legend, but some judged that in the case of ancient Israel, the record is more careful than that, and we do not have sufficient cause to doubt that this part is historical.

The distinguished materials scientist Sir Colin Humphreys has looked in detail at this and other parts of Joshua, and carried out archaeological

---

[6] Joshua 3:15–16.

and geographical studies of the area of Israel in question.[7] It turns out
that, far from it being impossible or unheard of, it has, in fact, occurred
many times that the Jordan River has abruptly stopped flowing,
owing to earthquakes in the region, which can cause a mud slide which
temporarily dams the river. This is consistent with the phrase in the
book of Joshua where it says that the water 'piled up in a heap'.
Furthermore, the location of this 'heap' of water is not nearby to where
the Israelites crossed, as one would expect if the account were legendary,
but 'at a great distance away, at a town called Adam in the vicinity of
Zarethan', *and this is precisely where the mud slides have regularly occurred.*

Taking all the evidence into account, Humphreys argues that we
can be open to the possibility that what is written in the book of Joshua,
in Chapter 3 and verses 15–16, is, at least so far as concerns the flow of
the river, a statement of what took place.[8] This illustrates the way that
scientific study can help in biblical study. It can help in determining
what is the literary genre and hence how to interpret the text. According
to Humphreys' interpretation, the river temporarily stopped flowing
owing to an earthquake and a mud slide. That is, the cause of the inter-
ruption could have been entirely within the natural course of events.
Nevertheless, the timing was remarkable. Some may want to regard the
timing as a coincidence with no larger meaning. This is an example of
the fact that no event carries its interpretation within itself in an incon-
trovertible way; it is always the case that we must come to our own
view of what each event shows and signifies.

This example opens the way to thinking about other cases.

> That very night the angel of the Lord set out and struck down one
> hundred and eighty-five units [eleph] in the camp of the Assyrians;
> when morning dawned, they were all dead bodies.[9]

What is reported here is that many Assyrian soldiers died. The number
given in many versions is eighty-five *thousand* but this is probably a mis-
translation since the Hebrew *eleph*, often translated 'thousand', commonly refers simply to some sort of grouping, such as a clan or troop.[10]

---

[7] C. J. Humphreys, *The Miracles of Exodus* (Harper, 2003).

[8] As we were completing this chapter a letter appeared in the prestigious journal
Science, asserting that 'There is no evidence for major Biblical events' and giving examples.
Science **358** (2017), 1142 (see also footnote 10).

[9] 2 Kings 19:35.

[10] A similar translation issue resolves some puzzles about the numbers of people
involved in the Exodus from Egypt. Humphreys argues that a correct interpretation of
the Hebrew text of Numbers gives about twenty thousand people, not two million as

So the number was not 185,000, but it was large and the effect was clear: the army withdrew and Jerusalem was saved. The Jewish historian Josephus, writing very much later, quotes Berossus, a Babylonian historian, as saying that a 'pestilence' broke out in the army camp. The interpretation of what was going on, says the biblical text, was God's action to protect Jerusalem. The physical cause, we may suspect, was an outbreak of cholera or some other disease. If this is so, then, like the case of the river Jordan, a natural process (earthquake, outbreak of disease) has been interpreted from the point of view of gratitude for help.

The sense of relief amongst the Jewish population was, one might add, perfectly understandable. The Assyrian soldiers were part of a brutal army intent on slaughtering or enslaving the inhabitants of Jerusalem and neighbouring towns and villages.

There is a danger in assuming that the function of the Bible is to hold all generations forever to the specific degree of understanding which its authors had, when it comes to the processes of the natural world. In the Hebrew Bible the distinction between a natural process and a process enacted directly by God is not sharply drawn. The Hebrew language sometimes begins to draw this distinction, in a subtle way, by the use of a variety of names for God. Some names such as Elohim (*God*) hint at the idea of a unity behind what we might call the forces of nature; others such as Hashem (*the Name*) are more intimate. These distinctions may be lost in translation unless we learn to look for them. The important aspect of the biblical account is not its limited grasp of the causes of disease, but its gratitude and desire to learn. That is what we should allow to mould our thinking.

By the time we come to Jesus, he himself, and other rabbis, were much more hesitant about saying which natural event happened for which reason. Commenting on the tragic deaths of eighteen people when the Tower of Siloam fell on them (Luke 13), Jesus was reluctant to discuss whether they deserved it more than anyone else. Similarly, when he was challenged about the causes of congenital conditions (John 9), he was more concerned about how to help. Every event can be co-opted by us so as to show God's glory somehow or other, whether by living graciously because of it, or by living graciously despite it.

stated in many translations (C. J. Humphreys, 'The Number of People in the Exodus from Egypt: Decoding Mathematically the Very Large Numbers in Numbers I and XXVI', *Vetus Testamentum* **48** (1998), 196–213).

A related danger can lie in too quickly supposing that we can interpret natural processes or historical events in our own day in a way similar to that on show in the Old Testament. The vast majority of such attempts have been misconceived. They have usually amounted to an attempt to support the speaker's own agenda or opinions.

19.3 Natural processes

We discuss in various parts of the book our view that it is not correct to make a sharp distinction between God's creative act and the outworking of a creative natural process. (In the case of a destructive process, more caution is needed, as we explained earlier.) This is also the view implied by Jesus when he spoke of God 'sending the rain on the just and the unjust alike'.[11]

Jesus' statement can be seen as follows: rain falls impartially, without regard to the moral qualities of people in need of water, and this is something we could usefully reflect on. If we know God's nature correctly, then we will know that this fact about rain can be used as an illustration. It illustrates the fact that God does not calibrate His generosity according to the attitudes of the recipient, and we should do likewise: we should act well towards those who seek to harm us. This is the point of Jesus' statement, and its truth does not rely on, or require, any particular model of the water cycle and the processes of evaporation, cooling, condensation, and gravity. Rather, what it requires is a sound general sense of God's character, and an awareness that the natural world is the vehicle of God's relationship to us, and ours to God.

In a similar way, the psalmist was not entirely ignorant of the ordinary processes of nature. He knew that some animals are herbivores and others are carnivores, that plants stay in one place but animals walk about, that water is central to life, and so on. Psalm 104 is a lengthy reflection on the relationship between God and the natural world, a 'reading' of what is going on. Even when it was first composed it would have been accessible to people outside the Jewish faith, because of its resonance with *The Great Hymn to the Aten*, possibly written by Akhentaten himself when he was Pharaoh of Egypt in fourteenth century BC.[12]

---

[11] Matthew 5:45.
[12] Wagner and Briggs, *The Penultimate Curiosity*, 357.

Bless the Lord, O my soul.
    O Lord my God, you are very great.
You are clothed with honour and majesty,
    wrapped in light as with a garment.
You stretch out the heavens like a tent,
    you set the beams of your chambers on the waters,
you make the clouds your chariot,
    you ride on the wings of the wind,
you make the winds your messengers,
    fire and flame your ministers.

You set the earth on its foundations,
    so that it shall never be shaken.
You cover it with the deep as with a garment;
    the waters stood above the mountains.
At your rebuke they flee;
    at the sound of your thunder they take to flight.
They rose up to the mountains, ran down to the valleys
    to the place that you appointed for them.
You set a boundary that they may not pass,
    so that they might not again cover the earth.

You make springs gush forth in the valleys;
    they flow between the hills,
giving drink to every wild animal;
    the wild asses quench their thirst.
By the streams the birds of the air have their habitation;
    they sing among the branches.
From your lofty abode you water the mountains;
    the earth is satisfied with the fruit of your work.

You cause the grass to grow for the cattle,
    and plants for people to use,
to bring forth food from the earth,
    and wine to gladden the human heart,
oil to make the face shine,
    and bread to strengthen the human heart.
The trees of the Lord are watered abundantly,
    the cedars of Lebanon that he planted.
In them the birds build their nests;
    the stork has its home in the fir trees.
The high mountains are for the wild goats;
    the rocks are a refuge for the coneys.

You have made the moon to mark the seasons;
 the sun knows its time for setting.
You make darkness, and it is night,
 when all the animals of the forest come creeping out.
The young lions roar for their prey,
 seeking their food from God.
When the sun rises, they withdraw
 and lie down in their dens.
People go out to their work
 and to their labour until the evening.

O Lord, how manifold are your works!
 In wisdom you have made them all;
 the earth is full of your creatures.
Yonder is the sea, great and wide,
 creeping things innumerable are there,
 living things both small and great.
There go the ships,
 and Leviathan that you formed to sport in it.

These all look to you
 to give them their food in due season;
when you give to them, they gather it up;
 when you open your hand, they are filled with good things.
When you hide your face, they are dismayed;
 when you take away their breath, they die
 and return to their dust.
When you send forth your spirit, they are created;
 and you renew the face of the ground.

May the glory of the Lord endure for ever;
 may the Lord rejoice in his works—
who looks on the earth and it trembles,
 who touches the mountains and they smoke.
I will sing to the Lord as long as I live;
 I will sing praise to my God while I have being.

May my meditation be pleasing to him,
 for I rejoice in the Lord.
Let sinners be consumed from the earth,
 and let the wicked be no more.
Bless the Lord, O my soul.
Praise the Lord!

Please don't let anyone take this type of poetry as some sort of alternative science! And don't take all the metaphorical language literally! Rather, let us pick up the sense of fascination with both the humblest and the most awe-inspiring processes of the world. Let's receive the thrill and marvel and close attention—the same combination which drives us to scientific study. Note also the sense of concern for ordinary creatures going about their lives, and the awareness of ordered routines in the processes of the world. The attitude expressed in the poem is one of celebration and respect, not fear. Finally, let us also receive the note of mourning and anger. When, in the New Testament, the apostle Paul wrote about sinners, he quickly added, 'and I am the chief one'. This is how to read statements about sinners in the Psalms: the sense of anger is directed not at the 'other' but at the forces and attitudes which damage us all and may find a foothold in any of us.

The poem is, to a large extent, a meditation on ways in which the natural world reflects aspects of God's character. In so far as the psalmist understands the latter correctly, he will be able to read the meaning on show in the natural world correctly.

Note the order of the logic here. One must know God's character correctly in order to interpret natural processes correctly, not the other way around. As we learn to do the former, we learn to recognize what aspects of the world can be celebrated and what should be opposed or changed. One does not go about this the way some commentators have done, just making up whatever story suited them, and backing it up by picking out whatever parts fit their version. For example, the operation of impersonal forces in the natural world provides both food and starvation to ordinary animals. These forces show neither bounty nor cruelty but indifference. It does not follow that God is indifferent, however. God may be as moved by His creation as we are, precisely because He has imparted to it the opportunity to be itself and not His mere puppet.

It follows from all this that there is every encouragement to find out about the natural process, and recognize that there is one, and this includes such processes as biological evolution. We can compare our interest in biological history with our interest in astronomy. A certain kind of marvel excites our eagerness to find out about the stars. And it turned out that, when we discovered what is the actual distance to the stars, then the marvel only increased a hundred-fold! A similar reaction applies to the study of biological ecosystems. The long, rich story of geological eras and mass extinctions and gradual diversification is more wonderful, not less wonderful, than a smaller miracle in the relatively recent past. The issue with an observation such as common ancestry (of humans and other animals) is not whether the bare physical fact is so, but what we think it implies about the value and nature of humans and other animals. There is no need to consider that it undermines the value of either. For this reason, and the reasons already given in Sections 19.1 and 19.2, we judge that the incomplete knowledge about physical things that is displayed in the Bible is not a problem.

We thus conclude that the difficulties in handling Scripture are centrally concerned with moral issues rather than descriptions of physical processes and states of affairs (i.e. science).

19.4 Moral issues

Although the Bible is often thought to be 'a book', it is in fact a collection of documents written at various times over a period of around eight hundred years, and the fact that it coheres together is itself rather remarkable. The main thing we would like to suggest is that it is correct to see these documents as the record of a journey, and a learning process, and a study. In a journey, you do not expect all the features of the destination to be present at the start. In a learning process, you see people learning; you don't see people already getting everything right at the beginning, and never making mistakes. You see *something gradually coming into focus*, as C. S. Lewis put it. Finally, in a study, you make notes as you go along. In an apt metaphor suggested by John Polkinghorne, the Bible might be loosely compared to a 'laboratory note book'.

These remarks may help us to see what it means to say that the Bible is inspired. We recognize the inspiration not in the sense of every sentence being perfect, but in the sense of the influence which made the journey possible, and which promoted the learning, and which gave the signals which were noted, however imperfectly. This, as we understand it, is the broad thrust of what Jesus was recognizing in his handling of, and response to, the Hebrew Scriptures.

Suppose we detected conscious life on another planet, and, to our dismay, it turned out that the aliens living there had banded themselves into tribes and these tribes went about trying to obliterate one another. Suppose the leaders and their followers thought that the best way to win renown, and a long life for themselves and their heirs, is to be the most accomplished at this fight for supremacy. Suppose you introduced yourself to such a tribe, and the tribe was quite impressed by you, and wanted to show what a great friend to you they were. Unfortunately, they will not be good at understanding your values. Before you know it, they might go on a killing spree in your 'honour', before you had managed to convince them not to behave like that. You can imagine that you are going to find it difficult to convey your mixed feelings when you are invited to the 'victory' celebration. They will probably write something pretty confused in their special book. But would you rewrite their book?

The celebration of tribal supremacy is practically a universal feature of human life before people slowly learned to think differently, and this story of the warring aliens is, in fact, the story of where all present-day human societies have come from. When it shows us warring tribes, the Bible shows us not someone else's history but our own history, the history of all of us. And it shows us who we would still be, if our ancestors had not paid attention.

It is through human life as muddled as this, through processes as humble and imperfect as this, that, it appears, God has to work. God does not rewrite anyone's story, but interprets for us. And in the end God offers us something better than a book.

We don't think the people speaking for God in the Bible always said good things or read events correctly. We are not meant to read it as a story of 'goodies' and 'baddies' in which the 'goodies' get everything right and never make mistakes. It is well known that leaders such as David were flawed; there is no shame in admitting that prophets such as Samuel were also flawed. In fact, it is our moral duty to see this.

We are meant to mull it over. Modern-day Jewish culture shows what mulling over the Hebrew Bible (what Christians call the Old Testament) actually leads to—an impressive, energetic, and civilized culture which prizes learning and entrepreneurship and has a lot of wisdom about living and a great sense of humour. Also, Jewish culture has developed a tremendously steadfast sense of *keeping going* in the middle of pain and difficulty, because they have loyally held on to the fact that nothing eludes or thwarts God, and *Yahweh Elohim* loves and treasures us even if no one else does.

To illustrate our approach to handling the Bible, we will comment on some of the passages where the difficulties of the Bible are most apparent. Of course, there are also many passages where the message is less risky and easier to receive.

19.4.1 Abraham and Isaac

Let us consider the story of Abraham and the binding of Isaac (Genesis 22). This is a dangerous and disturbing story, because in it God commands Abraham to kill his own son as a ritual offering, and then God stops Abraham actually doing it once it is clear that Abraham is ready to go ahead and do as he was told. The story is written without adornment, and consequently carries the kind of power that an unadorned statement has. It exists in a kind of primordial quietness, a zone where one must tread carefully. It dares to grapple with dangerous inclinations that stir in human hearts. Its meaning has been puzzled over ever since.

At first impression, this looks like a kind of test, and a totally horrifying and objectionable one at that. Will Abraham do the awful thing that God has put him up to?

On a more mature reading, one realizes that child sacrifice is the very issue that this story is expressly designed to confront. We know that child sacrifice is wrong. But how do we know that? Largely because we have learned it from others. And where did those others learn it from? To a large extent, the proper moral respect for children has been learned from the very community (the ancient Israelites) that wrote this story. And they learned it from God.

The story shows this very learning process in action. That is why it ought to be treasured and mulled over. It is one of the turning points in the history of the human race.[13] The Danish philosopher Søren Kierkegaard told and retold the story many times, each more emotionally powerful than the last, and was inspired by the story to make great sacrifices in his personal life.[14] The English composer Benjamin Britten wrote a simple but profound musical setting for tenor as Abraham and alto as Isaac, accompanied by piano, with God's words sung by the two voices in harmony.

---

[13] Jonathan Sacks, *The Great Partnership: God, Science, and the Search for Meaning* (Hodder and Stoughton, 2011).

[14] Søren Kierkegaard, *Fear and Trembling* (1843; English translation Penguin, 1985).

Throughout history, societies have often been willing to treat their children as mere instruments of their parents' will, or as disposable objects. That is how the ancient Romans regarded them, for example. The story of Abraham and Isaac is asking the question, Well what do we think about this? God asks the question: Are you ready to acknowledge that you do not own your children and have no right to put them to your own uses nor to make them the bearer of your dreams? Will you, in fact, give them totally to me, in the most complete way that you can? Abraham answers, on behalf of all of us, 'yes'. And then God replies: Now that we have that clear, then I will let you take care of your children, and note this very clearly; make it a central lesson of your community, recorded in a story of intense primordial power: *you may not harm your children.*

As Jonathan Sacks has put it pithily and well: the lesson is that parents are educators, not owners. Our children are free representatives of God's image in their own right, not our property, and we should not harm them. These are the major lessons of this turning point.

There is also an astonishing trust at work among all parties in this story. Abraham trusts that somehow Isaac will not be lost to him; God trusts that Abraham will not act precipitately; Isaac trusts, against all appearances, that his father will not harm him. And all these attitudes of trust are vindicated.

The above is one legitimate way of reading this story. One can see that it is a legitimate reading partly because the ancient Jewish community prided themselves on the fact that they knew child sacrifice was an abominable practice, when others did not know this, so they must have learned it somehow. Their community became one of the most respectful of children in the ancient world. The Jewish community continues to value education highly to this day.

We do not know if the above reading is the only way, or the best way to receive the story, but we do know that this is not to be dismissed as a primitive and dangerous legend from a superstitious age. It is more like the asking of a primeval question. In the modern world, few parents will literally kill their children,[15] but they may well exert pressure on

---

[15] In the UK in 2013 the mortality rate per million children aged 0–14 years due to assault and undetermined intent ranged from 3.5 in Scotland to 5.6 per million in Northern Ireland. The homicide rate per million children under 18 years was similar. S. Jütte *et al.*, *How Safe are our Children? The Most Comprehensive Overview of Child Protection in the UK* (NSPCC, 2015). Available at https://www.nspcc.org.uk/globalassets/documents/research-reports/how-safe-children-2015-report.pdf

them to fulfil some project of the parent's choosing. We still need to learn the lesson.

19.4.2 Punishing the children?

A phrase in the book of Exodus is in a very prominent position and has been widely latched onto, mistranslated, and misunderstood:

> for I the Lord your God am a jealous God, delivering the crooked-ness of the fathers on the children, on the third and the fourth gen-erations of those who reject me, but showing loving-kindness to thousands of generations to those who love me and keep my com-mandments.[16]

The expression 'delivering [*paqad*] the crookedness [*avon*]...on the children' sounds a lot like punishing one person for another's failure, which is unjust, but since God is not unjust then either the ancient writer got it wrong or that reading of His statement is wrong. To get clarity on this we don't make a snap judgement but look at what the world is like and at what others have said about it. The ancient Hebrews were bothered by the same uncertainty, so the book of Ezekiel takes pains to make it as plain as day that God assesses each person on the basis of their own actions, not those of their parents. In short, God does not punish people for wrongs committed by their parents or grandpar-ents. Both Jeremiah and Ezekiel denounce the popular saying that 'The fathers have eaten sour grapes, and the children's teeth are set on edge.'[17] Contrary to what people instinctively assume, God is not mainly concerned with punishment at all. The phrase from Exodus quoted above does not even mention punishment as such, but it does mention repercussions.

The world is full of repercussions in which children suffer for what their parents do. If a brutal parent strikes a child, then the child is struck: the parent's crooked act is directly delivered; its horrible consequence is pain for, and possibly brutalization of, the child. So this is also part of the truth about reality: no one, not even God, is going to change the law of cause and effect to prevent us hurting our children. So we had better watch out! What we do matters! God does not prevent the consequences of our choices, because such a prevention would destroy us.

---

[16] Exodus 20:5; the sense of 'jealous' here has the connotation of being involved, car-ing about the outcome, as opposed to being aloof or not bothered.

[17] Jeremiah 31:29; Ezekiel 18:2.

In the example we just gave, the direct impact of a crooked act is plain. The writer of Exodus has in mind a less direct but more important application to communal life. The context is a proclamation to the whole community of Israel (the announcement of the law, and its summary in the Ten Commandments). If a community—a generation of parents—rejects contentment, truth-speaking, straight-dealing, faithfulness, respect for life, wisdom, thankfulness, seriousness, and reverence, in favour of coveting, lying, stealing, adultery, murder, disrespect, thanklessness, flippancy, and pride, then the children and grandchildren of that community will find themselves in a miserable condition. And whether we like it or not, and whether we ascribe it to God or not, this is simply the truth. What is 'delivered' [*paqad*] onto the children is indeed not some punishment but the 'crookedness' [*avon*] of the previous generation. Not having seen truthfulness, they will struggle all the more to know what it is; not having seen respect for life, they will struggle to develop it.

Experimental research reveals to an extraordinary degree how even by the age of 2, children are learning from the interactions between grown-ups.[18] The writer of Exodus is dealing in frank and penetrating insights about communal life. It is the kind of truth that any community needs to hear.

The aspect that makes us uneasy is the language in which these cause-and-effect processes are attributed to God. An atheist could here simply part with the personal language and say 'dear friends, let's behave well, because if we don't then, by the inexorable working of cause and effect, our children will suffer for it.' The point of the personal language in the Bible is, ultimately, that the reality we spring from can and does have opinions about us, so the theist version is the correct one. It helps to know this because it indicates where we have to look to avoid getting ourselves into such awful cycles of pain and tragedy. We must take on board the right set of attitudes, not just the bit about lies, cheating, and murder, but also the bits about thankfulness, seriousness, and getting it into our heads that despite everything we are passionately treasured.

---

[18]   A. Waismeyer and A. N. Meltzoff, Learning to make things happen: Infants' observational learning of social and physical causal events. *Journal of Experimental Child Psychology* **162** (2017) 58–71.

19.4.3 The Exodus

We come now to the most difficult of our examples, and we do not claim to know what is best. We offer the following pointers in hopes of encouraging people with the view that there is nothing to be feared from handling the Bible in a grown-up way, allowing full range to the types of literary genre to be found in it.

Indeed, *freedom from fear* is a central issue here. Present-day followers of Jesus may feel themselves to be caught between two fears. One fear is that if we allow ourselves to question what is affirmed by any part of the Bible, then we will not know where to stop, and the whole Christian witness will unravel. The other fear is that if we affirm everything that each part of the Bible affirms, then we attribute to God attitudes and practices that we cannot square with compassion and justice. So people are caught between these powerful forces.

A doctrine of Scripture needs to pay attention to the broad brush of the testimony of the historic church, and a correct doctrine is not necessarily either the modern conservative one or the modern liberal one. We (the authors) judge that a correct attitude involves awareness of what both want to affirm and living with the tension. The evangelical or conservative approach affirms duty to respect, and learn directly from, Jesus himself, as recorded in the Gospels, and to work out from there to the rest of the Bible. The approach called 'liberal' affirms that loyalty to God is loyalty to truth, which implies a duty to take up and respect properly constituted intellectual enquiry into the sources and likely nature of the texts. Neither of these approaches denies what the other is affirming. Rather, they each act as a check on the other, insisting that neither consideration is neglected, and out of this tension the learning process is born. The tradition of the believing community can serve as a further critical faculty against which to benchmark each new approach.

This turns on what we think Jesus himself thought and taught. It seems to us that he had a rather thoughtful and nuanced view in which he is careful not to endorse everything that people think they can find in the Hebrew Scriptures, but he endorses what he considers to be the central message. He certainly did not endorse the attitude of the more conservative element of his day, with which he often clashed.

In each part of the Bible, our duty is to try to discern what type of literature we are dealing with, and to receive the account accordingly.

This process can be helped by more general studies of what sorts of literature the people in any given time and place were writing.

In the case of the Exodus account (the plagues in Egypt and escape of the Hebrew slaves), such studies strongly suggest that the genre is somewhere between history and proclaiming the character of God in a form of theological polemic; we are not able to tell exactly where to draw the boundaries. We find the account of the plagues disturbing, and one may legitimately suspect that Jesus would have found them disturbing too. Perhaps we are meant to find them disturbing. But certainly, in ordinary justice (and as the Bible strongly affirms elsewhere), the punishment should fit the crime, and it is impossible to make a case that killing the first of the next generation of the Egyptian families could be just, even for the crime of subjecting others to slavery. Yet that is what the writers have ascribed to God.

This has been much agonized over by commentators. It may be that the plagues were a sequence of events with natural causes, but even if this is so, they are presented to us as God's action against the nation of Egypt. Some say that God has the right to withdraw the life that He has given, which may be true but it is a highly risky way of speaking. We are not sure about it. In human life, if a giver asserts the right to withdraw a gift, then the gift was not fully given. Perhaps this is right: perhaps we are not fully given our life until we offer it back to God, because this is the only way it can then be fully given to us.

Whatever the answer to this may be, when reading of another's pain and grief, it is not just understandable but correct to have sympathy for their predicament. Therefore, when reading the account of the plagues in Egypt, our sympathy should extend to all people involved in the story.

A view that seems to us to do justice to the account, and to ancient literary practices, is the following.

In the account of the plagues of Egypt we have a form of religious writing in which the authors simply did not see the Egyptians as equally precious to God as themselves, and they are writing in a genre somewhat like modern-day Marvel Comics where it is legitimate for the superheroes to slaughter the 'aliens'. One can benefit from the account by temporarily switching off one's perfectly proper sympathy for real flesh-and-blood Egyptian families, and allowing 'Egypt' to stand in for 'cruelty' and 'superstition' and 'determined opposition to truth and justice'. If this is right, then the account is loosely historical but written without an interest in doing history the way a modern historian would.

Instead, the aim is to make it clear that the 'gods' of the Egyptians do not shape up, and that the liberation did not come easily but had to be fought for, by God on behalf of His people. However, having noted that, one must allow one's fellow-feeling for ordinary people, of whatever race or nationality, to reassert itself. This, surely, is the attitude that is allowed to call itself 'Christian', whereas an attitude that displays no such sensitivity has no right to that label.

The Exodus story has a very powerful image of the Jewish families killing their sacrificial lambs and daubing the doorposts with blood, waiting in trepidation for the bringer of death to pass over. This image is taken up in the Last Supper, which took place at the celebration of Passover, so we should by no means reduce its power or undermine it. This is why we are not confident of what should be said about the story of the Exodus. However, whatever is said must not be such as to harden the hearts of modern-day Christians. Had a modern-day representative of Jesus' attitudes approached Pharaoh on behalf of the people, one can easily imagine his preference for a non-violent protest, even at the cost of his own life. Indeed, it is conceivable that non-violent protest was exhibited by various people at the time, and they were summarily slaughtered by Egyptian soldiers. Such brave action, if it happened, has been lost to history.

In the time and place of the Exodus, no one was able to give the sort of sustained teaching and demonstration of world-changing truth that could capture the hearts of a community of peaceful followers. Instead there was a sequence of painful events and courageous determination from Jewish leaders.

We should respect the great passion for justice and sense of liberation from oppression which is the central theme of the story. It has, in fact, been a story which several liberation movements have turned to for encouragement in their struggle, and its powerful images have been central to that role. Therefore, we should honour its power and contemplate our complicity in unjust practices in our own era. Martin Luther King famously drew on the imagery of the Exodus in his inspired leadership of the struggle for civil rights in America:

> Like anybody, I would like to live—a long life. Longevity has its place. But I'm not concerned about that now. I just want to do God's will. And He's allowed me to go up to the mountain. And I've looked over. And I've seen the Promised Land. I may not get there with you. But I want you to know tonight, that we, as a people, will

get to the promised land! And so I'm happy, tonight. I'm not worried about anything. I'm not fearing any man! Mine eyes have seen the glory of the coming of the Lord.[19]

The Promised Land envisioned by Martin Luther King was a country free of racial prejudice, and he knew that it had to be championed actively; it would not come by sleepwalking. The power and the passion of 'Let my people go' cuts through the excuses and sustains the quest for genuine freedom and *shalom*.

19.4.4 The surprising teaching of Jesus

The cover of this book is a detail from a painting by Roger Wagner entitled *Writing in the Dust* where several figures are (appropriately to our title) straining to see something just out of sight. We have included the full picture inside the covers. In Roger's unmistakable style, interposing contemporary technology and eyewear in a first-century scene, he portrays an incident in which religious leaders bring to Jesus a woman who had been caught in adultery.[20] They remind him that the Mosaic Law demanded capital punishment by stoning, and invite his judgement. Before he answered them, Jesus bent down and wrote with his finger in the dust. We do not know what he wrote, but they pressed him for an answer, and when he stood up he invited anyone who was himself blameless to initiate the killing. One by one they slunk away. Jesus then turned to the woman and asked if anyone had condemned her. When she replies no, he says neither will he, but tells her to go and change her life.

Roger has written a poem to accompany the painting.[21] He imagines how, reading the story or looking at the picture, we are both hypocrite and sinner. The story in the Bible is sometimes given the heading 'The Woman Taken in Adultery'. Roger subtitles his poem 'The Men Taken in Hypocrisy'.

> The beating of a swallow's wings,
> A stone jar poised as if to fall,
> A fierce and unforgiving sun
> That beats upon a whitewashed wall

---

[19] Martin Luther King, I've Been to the Mountaintop, sermon delivered 3 April 1968, Mason Temple (Church of God in Christ Headquarters), Memphis, Tennessee.

[20] John 8:1–11.

[21] Roger Wagner, *The Nearer You Stand, The Farther Away* (forthcoming, 2019).

That finds and tracks each human flaw
And reads the writing in the dust
Of broken hopes and powdered dreams
And love reclassified as lust.

Where images of shameful death
Describe a life defined by blame,
The beatings of a swallow's wings
Above a place of public shame

Are like the barest breath of grace
That stir the unforgiving air:
That shift the gaze and lead the eye
Beyond the camera's fatal stare

To where one writes in grit and dust
—Of dry bones in a bone-dry place,
Of broken hopes and powdered dreams—
The unseen, unhoped, words of grace

Which free accuser and accused
Which spell out where that life begins;
A motion like a breath of grace:
The beating of a swallow's wings.

Christian reactions don't change the structure of the impersonal physical processes at work in any situation (including neuroscience, etc.), or deny that those processes are happening, but they do reject fatalism, and they hold open the possibility of rethinking what is really going on in any human situation, and what are the possible responses open to us.

The surprises in the accounts of the miracles of Jesus arise because those seem to go against the normal pattern of behaviour of the material world. There are plenty of things that Jesus did and said that present no apparent departure from any scientific regularity, but which are utterly astonishing in their challenge to accepted human norms. Perhaps one reason why they do not surprise us more is because we are beneficiaries of changes in attitude whose origins can be traced to Jesus.

In our glossary at the end of Chapter 3, we gave our operational description of Christianity as learning from Jesus *Christos* and the community of his followers and coming to know him in response. This carries the risk of being misunderstood as taking Jesus as simply a good moral teacher. His moral teaching, in action and in word, is indeed astonishing. To use a recently rediscovered word, it awhapes. But you cannot leave it there. We have noted that science is possible because the world is intelligible. We have tried to describe how this accompanies rather than negates the sense that the material world is a window which we can look through as well as at, and thus we are drawn to something beyond. Even more so, as we learn from Jesus and the full implications of his life and teaching, we are drawn to someone infinitely greater.

20

A conversation about the themes, continued

This is the continuation of the conversation whose first part forms Chapter 2. The conversation was a single whole, but we have broken it up here in order to help the reader follow the train of thought. The discussion was broadly shaped by the themes of the book without following them slavishly.

20.1 Third theme: uncertainty

HH: The way you're describing this resonates strongly with my feelings toward the issue. But I actually quite frequently have people say, when they are comfortable enough to ask me this level of personal question, they often say: I hear everything you're saying, it's quite reasonable. But I don't understand why you are not an agnostic. They say, I hear everything you say about uncertainty, about the arguments not being compelling, and stuff about being a human and how things are just so unclear and just muddled; you're just curious and see things. So why are you not an agnostic? Andrew, how would you answer that question? Why are you not an agnostic?

AS: I am going to say something brief and then I'm going to stop and come back. Why am I not an agnostic? It is because God loves me and I begin to feel that I begin to understand what it means to love God.

AB: I think that's right. I think you can only enjoy the relationship with a level of commitment that I think it's hard for the agnostic to muster. But I suppose the other thing, just standing back from the relationship issue (this is important in the book also when we're talking about science), is to realize that there's no one thing called uncertainty and there's no fixed percentage of uncertainty or something like that. It's much more to do with judgment.

Of course, sometimes in a scientific experiment or calculation you do precisely quantify the uncertainty. But here we're talking about a different thing, which is that there are degrees of confidence and degrees of uncertainty that are appropriate in different situations. So, if we just talk about the science for a moment, on the one hand we want to say that when science is conducted, it can only be conducted in a spirit of humble open-minded inquiry, where you're doing the research precisely because you don't know everything and you want to find out more. Nevertheless, it's also the case that there are some bits of science that we're pretty sure of. We are sufficiently confident about the phase diagrams of aluminium alloys to literally stake our lives on them every time we take a flight in a plane. There are some areas of science that we're very confident about and there are others that it's harder to be confident about or where the uncertainty is greater. And I think there's no shortcut to this. I think it is a matter of professional judgment and expertise that's born of working at these things, and it is part of the apprenticeship that goes into being a scientist. You have some tension there because, on the one hand with science we do want to engage everybody and make sure that we learn from different points of view and that everybody is part of the debate. Nevertheless, there are times when you have to say: this is a matter of professional expertise and these are things that we can be reasonably sure of. There are different degrees of uncertainty like that. To some extent, for me and Andrew it's part of our professional responsibilities as professors at Oxford to help the next generation see where the appropriate places of uncertainty are, where they might get a fruitful Ph.D., or do successful experiments or calculations. So, if that's true in science, I think it's also true in the life of faith. There are different kinds of uncertainty that are appropriate to different kinds of questions. And these questions and uncertainties are not incompatible with the confidence in the relationship with God, even though we recognize that that's something where we're learning and growing and finding out more.

HH: I want to follow on to that by pressing on the question of the analogy or disanalogy, probably I should say in plural analogies and disanalogies, between certainty in science and in the life of faith. Let me just start by prefacing that I think there have been two extreme positions over the history of Western society in the past few hundred years. On the one hand, there have been those who treat science as achieving a very high level of certainty and religion as this kind of mushy, subjective, believe-what-you-want-to attitude. Actually, in the United States, it's almost as if society has encoded that, so that religion is a personal thing. There is no certainty, it's not objective. Science is just the opposite.

There have been some movements towards the opposite extreme. You can name the twentieth-century philosopher called Paul Feyerabend, who claimed that science, if you really looked at it, is completely chaotic and there is just almost no certainty in science. And so, I am curious from your point of view, being workers on the ground, what is your experience? Do you feel like there is a radical difference in the intellectual grip you have in your scientific work versus in your life of faith? And if there are disanalogies, what are the key points where you feel there are differences between these two modes of thinking?

AB: I certainly found differences, perhaps through my experience of having studied at university degree level both theology and physics. I found one difference to be this way, that in physics I get almost a serotonin rush from 'getting' something. It could be a concept that had previously eluded me that I now feel like I got my head around and it clicks and makes sense to me. It could be proving something I was trying to prove, where you really can prove it—say, something in Euclidean geometry. Or you are doing a calculation and you get a new level of confidence that this is the right answer. Or it can be making something work: you've got a piece of apparatus or an experiment or something that didn't work and now it does work and there's a great satisfaction in that.

When I came to study theology that sort of thing eluded me, at least in academic theology. You don't get to a point where you've solved a problem in academic theology. It doesn't work like that. That to me is a big difference. It's one that academically and emotionally I found quite hard when I was studying theology.

AS: Sometimes in science I enjoy the feeling of 'I get it' and then with God it is a delicate feeling of 'I have been got'. But I have found, several times, that after lengthy puzzling over some religious question, I finally arrived at a way of seeing it that seemed right to me, and then, afterwards, I found that other people that I respected were saying that very thing. This has happened several times, and it gave me a sense that we were onto something objective.

Also I would like to add some other thoughts and I can't easily get them all marshalled. Let me say first of all: insofar as there may be some radical views along the lines that science is itself radically chaotic, I don't buy that. I have a pragmatic response to that. A bit like Andrew saying, I know the properties of aluminium and I'm willing to place my life in hands of the aeroplane. That's reliable enough for me. Next I wanted to think a little bit more about the question of agnosticism. So, coming back to the question about agnosticism: to me, a lot of this is about one's feeling oneself under a duty. Duty is a big part of my sense of who I am and what I am for. To me, it's not a question of 'Oh well, we all just come to our own opinions about religious questions.' No, we are under an obligation to come to true opinions insofar as we can. I think that what happened to me as a younger person (not very young: a young adult) is like getting to the place where I felt it was actually my duty to acknowledge that what Jesus was showing us demands a response from me. One had to humble oneself and acknowledge that one had been in the wrong. And now one was willing to place one's allegiance differently. One had been in the wrong about very many basic things about what the world is, our place and our part in it. But that one was willing to go in a better direction. Having made that decision, that doesn't automatically settle everything. You'd never really think, 'Oh great, I'm now in the right place, so everything's all sorted.' But what it did do, for me at least, what it gave me, was a tremendous impetus to check it out. When you are actually treading onto a bridge that you have been unsure of, you pay extra attention to whether it is secure, not less. So one looks hard at all those questions which still remain with one. There is a tremendous impetus to really give a lot of time and energy to that.

I suspect, for the most part, when my agnostic friend says to me, 'Aren't you really agnostic?', the truth is probably that I spent more time and energy thinking this over than the agnostic did, and I really have worked hard at this and I feel myself under a duty to continue with it. I just don't have the right to throw it over.

One of the questions one looks at of course is how it pans out in actual expressions of human life, in the many different cultural settings, in many different strands of historical development. There's a lot of things which one pays attention to. And the picture is by no means rosy. It's a mixed picture. But, nevertheless, you give due consideration to all these things.

So, to conclude, that's a large part of my response to why I'm not agnostic. It is because I can only do what I feel myself under an obligation of duty to do. I mean duty to truthfulness, to what I found when I looked. My duty to be honest and stick with what I have taken the trouble to explore carefully and not thrown over for insufficient reason. Now let me connect this with the experience of being a scientist. I think that there one is also under an obligation to behave properly and do science correctly but on the whole it's just a lot easier. It's like the difference between lessons and playtime at school.

AB: There are plenty of activities that we engage in where commitment is a prerequisite for experiencing the activity and finding out more about it. You can't enjoy a swim unless you dive into the cold water. It's not possible. I fly aeroplanes. You can't fly the aeroplane without committing yourself to taking off on the runway. It's a necessary part of the activity.

HH: Wouldn't you say also our society has a false stereotype of science not involving faith commitment, but surely science involves commitment?

AB: It's hard to think of any profession that involves a greater commitment. If you look at the commitment that's necessary even after the first degree: the commitment of doing a Ph.D. when you don't get paid very much and you work very long hours. After that, people have an extended period of post-doctoral employment on fixed term contracts with no certainty at all of the eventual career outcome. In fact, quite the opposite, because the probability of ending up in a tenured university post is not very great as a fraction of the number of postdocs. The last thing I want to suggest would be that that's the only worthwhile career. But what I'm saying is that for those who want to go into a career in research the required level of commitment is huge.

AS: Yes, I agree with that, and that it's an important human and genuine consideration. But I think Hans is also wanting us to reflect on the fact that even the very ideas that we write about in our attempts to publish scientific knowledge involve a commitment of ourselves. I've written about this elsewhere but I think it's a good moment to say it again here that whenever I've written a scientific paper, there's never been a case when I wasn't somewhat worried that I'd missed something. Every time I was worried that it would prove to be a case where basically I just got something a bit wrong, or that particular contribution doesn't really amount to much. So there is this risk that you might have made an error, or simply that you did something that's not wrong but it's not useful. That's the other way in which a contribution can fall by the wayside. A big aspect of science is that you don't go forward on the basis that at each stage you know for sure that what you're about to announce in your publication is certainly fine. You don't know that, you never know that. And furthermore, whenever there's a larger step made in science, a bigger contribution to knowledge, or something which is a whole new way of looking at things, then it's even more so that there is a need for commitment. You have to trust in your own ability to tell what works and what doesn't. The big theories that really make progress are especially like this. You can think of Harvey and the circulation of the blood and Einstein with general relativity, and Darwin and the other contributors with natural selection and so on. Each time the person who is really promoting the idea and putting it forward can't know for sure that they've got it right.

AB: Indeed, with quantum theory, Einstein for many years, possibly the whole of his life, was convinced that it was incomplete.

AS: Indeed, so Bohr and Bohm and everyone had to keep worrying at it. The point here is that scientists, in producing some of their best work, in fact, go out on a limb. And that's great. That is part of the creative process. To go out on a limb like that is somewhat like what we do when we talk about the word faith, when we use the word faith correctly.

HH: But, Andrew, isn't that now very different from the popular perception of science where people would suppose that you're talking about things that are established, I was going say, beyond doubt? I think that may often be a perception.

AS: I think the important difference may be that it's because science deals in those parts of knowledge that we can more confidently test and come to agreement on quickly. Therefore, even if initially there is this large contribution of faith, it's easier to then resolve any uncertainty and it happens more quickly.

AB: There are variations even within different academic disciplines in a university. There are different degrees of consensus that it's reasonable to expect. In the sciences, even though you know that at the cutting edge there may be disagreements, nevertheless, as time goes by you expect those to be resolved. You expect there to be a consensus that will emerge. But in some of the other disciplines it wouldn't be reasonable to expect a consensus; it won't happen.

AS: Yes, I'm just pausing to think of an example. I think that every discipline of human endeavour is seeking some sort of sense that we can agree on what is progress.

AB: You wouldn't expect ever, I don't think, to get complete consensus in a faculty of English literature about the interpretation of any particular work of Shakespeare. What about you, Hans? What's it like in philosophy?

HH: Oh it's all horrible, there is no consensus whatsoever. In fact, there is anti-consensus because it's one path to professional advancement for people to have different ideas than everyone else. Of course, there are various academic games to be played.

AB: Yes, but I mean from our point about different degrees and kinds of uncertainty in different areas. Nevertheless, it is also true that an undergraduate paper in philosophy could be just wrong.

HH: Yeah that's possible too. Sure.

AB: And every professor would agree that it's wrong.

HH: Yes, I agree. There is a sort of factual core about what a philosopher would say.

20.2 Fourth theme: freedom

HH: There's another big question that's been looming in my mind that I want to make sure we get in. It's a question with two sides. It's whether you have a final word for a certain reader.

Here are the two sides. The first side to the question is more generated out of the American situation. The American situation is that we have a very large population of confessing Christians, people who say they are Christians, and there is a strong anti-correlation with interest in science which is incredibly disheartening and discouraging for those of us living here. I'm wondering what your word is to Christians on why they should engage with science. I am talking about not necessarily Christian professionals who are already scientists, but as a young person, why should a young person who is already in college who was already committed Christian or feels drawn that way, why should they be engaged with science?

AS: Because science is just the name we use for part of our response to God. It's both a response to God, because it's a creative use of the faculties which He has furnished, and it's also a big part of how, in fact, we go about joining in with God in order to shepherd the world forward. Both the world of human relationships and also the wider natural world, insofar as we have a role to oversee the world. Science is a big part of our whole life. And it seems extraordinary to me that anyone would be in any doubt about that. I find it hard to know what to say. My positive feeling about science is so strong. How could it be in any doubt?

HH: I think, trying to put my finger to the cultural pulse, there is this other-worldly, almost Platonic attitude that this world is passing away. On one side, that is just bad theology to think this world is somehow an illusion and the true reality is somehow elsewhere.

AS: Yes. Let's come back to that in a minute. I agree with that but I want to allow Andrew to speak.

AB: There's a curiosity-driven science and there's a technological approach to it. First, the curiosity-driven science. One of my heroes, one of the people who I think deserves to be much more famous and celebrated than he is, is James Clerk Maxwell to whom we owe the whole unification of electromagnetic theory and light. In a sense, all of our electrical engineering is based on his equations. He was a person of strong Christian faith. He was the founding professor of the Cavendish Laboratory at Cambridge and, on the doors, carved a verse from Psalm 119:2, 'the works of the Lord are great, studied by everybody who has pleasure in them', and I think that's a pretty good starting point, actually. If you asked, what would the message be to a Christian believer in the USA, it could be this: If you believe the Psalms and you take them seriously, that's what science is. It is studying how God makes the world work. It's what Maxwell did and that's what every other scientist does whether or not they acknowledge that.

But there's also the technological aspect which is that the person who most glorifies God is the person who's most alive, and is alive in doing things and loving people. And our capacity to love people is enhanced technologically. You want to help people get better, so you use the best science of medicine that's available. You want to help people communicate when they're spatially separated, then you'd better use the best digital communications technology that's available to you. And these technologies are almost invariably based on discoveries that are made through what we call science. The two, of course, are gloriously entangled with each other. And so, if you're going to love people to your utmost, you're going to develop technology and use technology and if you're scientifically minded or if you're engineering-minded then you'll contribute to that.

AS: Let me add the thought that when Jesus tried to help his hearers, and therefore us, to get the big ideas he was trying to get across, we often see that he is drawing lots of different pictures and he can't just give it all out nice and clear and here it all is on the dotted line. Why is that? It's because his hearers then, and ourselves now, struggle to get into the mind-frame that he's offering us. We have to learn to see in these different ways. Of course, his big phrase is the kingdom of God. This is a large part of my reaction to the young Christian saying: so why take an interest in science? It's because science is right there: part of the kingdom of God. The kingdom of God is not some other-worldly thing that you go to after you die. It's the way of life that God encourages us to be living here and now, making full use of the riches that are available through the natural world and through our ability to interact with our minds and to love Him with the whole of who we are. Science is part of that. It's a big part. It's not some sort of alternative thing over there on the side that we do when we wander away from the kingdom of God. It's a big player right in the kingdom of God. This is part of what I want to add. I think I'd also say a couple of thoughts about helping people in the here and now. To have your heart set on things above means you're really wanting to share God's priorities and to be as He would have you be. And does that not translate into, therefore, if there is a large incidence of a disease such as malaria in parts of Africa, then, goodness, that matters to us. And we're going to do whatever it takes to address it. I'm reminded of Jesus' phrase about the sheep and the goats and he said to some: I was in prison and you visited me and I was hungry, you fed me. He's saying to us: 'When you do that, it's because you cared, you were interested in the world here and now and how people are going on and it mattered to you. Good for you. Come on in, that's what I'm all about.' This is a big part of what I think genuine discipleship of Jesus is in fact about.

HH: I agree with everything you've said. I think it can be quite effective with the kind of person I'm thinking of here. One other thing that

I think was included in some sense in what you said, but I'm going to ask you to unfold a bit more, is: what about my spiritual growth? What about me as a follower of Jesus? Do you have reflections from your own experience? How has science and your life in science in any way improved your life of faith, other than just your thinking, 'I did something that was good for the world.' Have you found that it actually helped you mature spiritually in certain ways, to do scientific work?

AB: Let me try two different kinds of answer to that question. One of them is very simply that if this world is God's creation, then part of knowing Him and appreciating Him is using all the resources available to understand and enjoy his creation. And I suppose the analogy there is with anyone who does something creative. It could be a painter, it could be Rembrandt. You could learn an awful lot about Rembrandt. You could read his biography, you could find out where he was born, who his parents were, who he studied under, how long he spent in Amsterdam, who his customers were, how his painting changed over his life and became less commercial and more interested in painting where his passions lay and so on. And you might, in principle, have exhaustive knowledge about that, if such a thing is possible. But if you've never looked at his paintings, if you never sat and studied *The Return of the Prodigal Son* and drank it in, you'd be missing something. You'd be missing something central to friendship with Rembrandt. You can think the same thing about composers, about Beethoven. It's not a perfect analogy, of course, because God's creativity extends beyond the material world and His concerns extend beyond the material world, so I don't want to overpress the analogy, but nevertheless, for me, there's something in that analogy.

The other answer consists of a story from Ghana. I was out in Accra and we were talking about some ways that scientists might contribute to the life of the Christian Church in Ghana and present in the discussion was a Methodist bishop. He told us about a situation that was uppermost in his mind at that moment. One of the pastors for whom he was responsible served in a northern area of Ghana, which is the poorer part of Ghana. There was a village that had been given a well by an international agency. Hitherto they had to go to the river and collect the water. I'm afraid it was the women who had to go to the river to collect the water. And they did, until the well came along, with clean, safe water for the first time in this

village. Shortly after the well was installed, there was an outbreak
of disease in the village. The local witch doctor persuaded the village that the disease was an expression of the wrath of the god of
the river because they were no longer collecting the water from the
river. And they must never again take water from this well. They
must revert to collecting river water and bringing it to the village.
The Methodist bishop had a question for us, and his question was
very simple: How can scientists help the pastor in that situation?
I would be very sorry if the answer came: I don't think science is
useful here; we just want to help people. I think quite the opposite.
We should bring to bear the resources of understanding of those
things, such as how the disease was caused (and I hope could be
cured), in order to help the pastor to enable those people to benefit
from safe water that doesn't have to be collected by women carrying it some distance from the river.

HH: It sounds to me like you were developing a case that, in a sense,
science is a very effective tool for combating sorts of idolatry, because
as science identifies and isolates physical things it explains that they
are not divine, they are just bits of mechanism in this network, then
our mind or worship can turn elsewhere. So science has been per-
forming that service for ages.

AS: I'd like to immediately agree with that perception of what we are
learning from Andrew's example, and I'd also like to mention two
other things. One is a difficulty; one is an encouragement.

The question from Hans was about the person who is thinking,
'I feel that my duty does draw me to follow Christ but I'm uneasy
about this scientific avenue—will it impact on me and cause me
maybe difficulty in a spiritual sense or even tempt me in to ways
I shouldn't go?' I want to say that my experience of that is that there
has been difficulty, but that isn't the reason not to do it, not to go to
science if that's where you can contribute. It's not all difficulty; there
are both difficult parts and rewarding parts. Andrew's spoken mostly
about the way it is rewarding and I completely agree. But let's agree
also there are these difficulties and what are they? They're social, cultural, mostly. There is, in our modern culture, a fairly widespread
behaviour in which Christianity is bagged in with a whole range of
other things which people disagree with, and then disdained. But I
think it's a disdain based on a refusal to really bother to understand
what it really is. This is part of our cultural milieu. I'd have to say

that sort of attitude can be found in the scientific community. We're not all completely immune to these sorts of ways of thinking. You have to acknowledge that that is so; therefore, you do feel at times as if you can't just 'come out' as it were and simply express the fullness of who you are, because you fear that it will be frowned upon and dismissed. And so you hold back and that is a spiritually difficult experience. That is an aspect of it all. However, on the side of encouragement, I would say that nevertheless when I look at the people who have gone before me in science and look at the intersection between those who are really competent in science and those who have chosen to express their faith in Jesus, I find that is a very interesting intersection. There are some very impressive people there, who do not disdain and are thoughtful about their commitment to Christ. And this does not at all result in there being something compromised in their science, not at all, not a bit. I've had much encouragement from that intersection of people; those who are both very competent scientists and who've gone before me on the way of Christ and said, 'This is a good way to go.' That's the encouragement.

HH: Yes. Now in another direction, our book is not intended to be a work of apologetics. But I wonder if we agree, it's very hard to put the right words to this, I have this sense and I wonder if you agree with it, that there really is a way in which everything we know of science fits better with a story that involves God the creator and involves the ethical duties that God the creator gives to us, and so and so forth. So in one sense there is a slightly apologetic function in the book because we're putting forward an indirect argument that science in this framework looks healthier and better to us than science detached from it. I hesitate to use the word argument because it's not a premised conclusion but we're in a way saying, 'If you take science very seriously, you probably should take God more seriously as well.' Is there something to that, do you think?

AS: I think there is, but I don't think we've chosen to write about that in this particular contribution except perhaps indirectly. There is Alvin Plantinga's book[1] in which he looks at the philosophical basis for why science is trustworthy. That's an example of someone trying

---

[1] Alvin Plantinga, *Where the Conflict Really Lies: Science, Religion, and Naturalism* (Oxford University Press, 2012).

to ask questions about how does science really fit into the context of a larger framework of how we discern truth. That's not the only thing you're referring to, I think, but yes there is a very positive story to be told here. Also, in another book of mine[2] I discussed the risk that when science is not well placed in a larger picture, then it risks becoming an object by which people exert power over others, or come to unjust or inhumane conclusions. Christian faith provides the larger framework that liberates science to become a good. But that's another story.

HH: I should have emphasized I know of Plantinga's book and I think this is a great contribution. I think though that I'm also more focused on the question for individuals, not really big abstract questions, but simply a lived life. I see on the one side the Christians for whom I think their lives would be enriched by deeper engagement with science. But I also think I see people who are committed from the start to a scientific life, and who see it as somewhat incompatible with a religious life. But I think our book will send the message that: no, in fact their life of science might be further enriched by letting religious commitment grow inside of it.

AS: Can I offer an expression that I found helpful in this area? I don't remember the name of the person who wrote it, but it was along the lines of 'We do not so much need to persuade people to be religious as to show them that they are religious.' This might help here.

Let's recall that we've acknowledged in the book that the word religion and the adjective religious can have a bit of a range of meanings. But we're here speaking about the more positive sense of those words, where they don't mean necessarily the effort to make ourselves acceptable and that sort of thing; they mean the willingness to recognize that there is meaning in our lives to be reached for and explored and religion is about that.

The person who's very serious about science will probably find science to be a very fulfilling aspect of their life. I think that they might be willing to acknowledge that they already are experiencing a sense of meaning in the contribution they are making. But this sense is open-ended. It is not completely addressed by finding the physical mechanisms of how things work. Nor even is it addressed by the larger or deeper ideas that science does offer, the big paradigms such

[2] Andrew Steane, *Faithful to Science* (Oxford University Press, 2014).

as variation and natural selection or the paradigm of operators and quantum fields. We sense these to be quite deep ideas and yet they don't actually tell us who we are. We want to know. And we don't want to know merely to satisfy our own curiosity either. I think we want to know in order that we should live, in order that we should give, and connect, and in order that we should not pass by what is really going on in the world.

HH: I've seen thoughtful secular authors who are writing in the genre of science and society saying similar things, although they feel a need to put at the end a disclaimer that this is not religious, this is not theism. There was one that we thought about before, *The Big Picture*, the book by Sean Carroll where he pushes very strongly in the direction, saying this is not just naturalism, but poetic naturalism. That it's supposed to include all the human things and then of course, but God is not there. A similar effort has been undertaken by a philosopher at Columbia University called Philip Kitcher, who wrote a book called *Living with Darwin*. He talks about how there is a need for some sort of new religion which doesn't involve God. Then he says many things that are quite similar to the things we say, but slightly different.

AB: There's also been a number of sociological studies. I think of one in particular as I speak where they've asked scientists about their practices and beliefs and what they find is rather uniform, at least in the US. There is a significantly higher number of scientists who would describe themselves as spiritual, than there are those who would say. 'I regularly go to church or I regularly engage in religious practices.' Many of these people who say, 'I don't engage in religious activities', nevertheless, are very comfortable in describing themselves as spiritual people. In terms of what is the added value of belief in God for the scientist, I think it's the complement of what I was saying earlier that if God is the creator and He created the world then there's a pleasure and enjoyment in studying His creation. If you take any creative work, a painting or something, you can experience a pleasure in the painting regardless of any knowledge or relationship with the painter. It's just there and it's enjoyable and it's pleasurable and it may give you new insights and experiences.

But now you can add a further dimension. Someone who's been a friend for over half my life is a very distinguished painter who lives in

Oxford called Roger Wagner. We've written a book together.[3] And I actually came to his paintings after I knew him. I already knew him; we'd already talked quite a lot about his thoughts and our shared ideas and so on, things that we enjoyed together. And then I came to these paintings. Now I can have a pleasure in his paintings just as anybody else can. But, for me, there's a whole extra dimension of pleasure because it's in the context of the friendship with Roger. Now, there's a pleasure in a scientific discovery or in learning something in science that you didn't know before, or perhaps that nobody knew before, or pleasure in getting an experiment to work or whatever. These are genuine pleasures that can be experienced by any good scientist, I think. But if you follow the analogy with my friendship with Roger, if that pleasure is in the context of the relationship with the Creator, then for me certainly it's enhanced and I suspect for many other people too.

HH: I think that's a wonderful new metaphor to combat the metaphor of Richard Dawkins saying about why can't you just enjoy the garden. This metaphor is much better because when we talk about God, it's not like we have a painting and then we want to add another element in the painting. We're not looking for a hidden creature; we're saying, 'Take the whole painting as it is. Don't add to it, but think of it as the work of someone with whom we can relate as a person.' And that's a very different metaphor. We're not directly adding to our catalogue of ontology. The thing about the Dawkins quote that bothers me is that it sounds like it makes theism into just one more scientific theory. There's the theory of the garden, then there's theism which adds in some fairies at the bottom.

AB: That goes back to one of our very first themes, of how we think of the physical world and we're not intending to add God as part of that, nor even as something like an extension. That isn't how Moses was invited to think, and it is not how Jesus approached it.

AS: Yes. We don't hear that from the Gospel narratives, and we don't hear that from the extensive theological reflection that exists in our community. At this point the Christian reader is thinking, 'Hang on a minute, what are you saying there?' And the other readers are thinking, 'It's all a bit puzzling now.' I think I'd like to welcome that

---

[3] Roger Wagner and Andrew Briggs, *The Penultimate Curiosity: How Science Swims in the Slipstream of Ultimate Questions* (Oxford University Press, 2016).

thought because, puzzling as it may be, I think that's genuinely where I am and I think where we all are. Let me return to it a little. You alluded just now to someone who had said that they wanted to have the Darwinian insights and so on, and some sort of new expression of religion, which would be without God. That returns me to the beginning of the discussion where I said, 'What is your word "God" there referring to?' If that person wanted to have a religion without a kindly king, supernatural figure, sitting alongside the universe, calling us to go to join with him in heaven, then good, I don't want that either, because that is not what I hear in Jesus' teaching, and no wonder. So I would perhaps invite that person to not think they already know, and experience for themselves what it really sounds like when you hear Jesus talking about God, if you take the attitude 'he is not talking about that fantasy everyone keeps squabbling about; he is standing on the same ground I am on and is talking about what he knows.' You're right to point out that Dawkins' picture of the garden plus fairies isn't how it actually operates.

HH: Coming back to Andrew Briggs, I found that metaphor quite a moving thought. I agree with that; you could find a painting, you could wonder out in the forest and find a painting sitting on the ground, and could think, 'This is just beautiful, I have no idea where it came from, maybe I don't care where it came from, I just study it for its own sake.' But then if you find out that this person you either already know or that you could come to know is the artist, it wouldn't add to your factual stock of information about what's in the painting, but it would help you understand. I mean, in one sense, I think the flow of information will go both ways. You would understand the painting in a way but also you would understand the person because of their work.

AB: Indeed.

AS: I think though we must move on to the thought that nevertheless the natural world is a dynamic world. It's not just a fixed thing like a painting; it's important that we are all playing a role in that dynamic.

HH: I think that we've found again and again that metaphors are great but we always have to stop at the end of the metaphor and say it's not perfect; here are some ways in which it might be misleading.

AS: If we are anything like talking truthfully, meaning if we've had the ability to understand things roughly right, then God is also playing a role in that dynamic beyond merely allowing it to just happen.

Maybe here is where the reader looking on from the outside begins to be nervous. Are these guys about to say God did this or that and exactly how do we speak that way? We have to tread carefully but we do affirm that as we read the picture of the natural world and try to discern in it, or respond to it, as one would to work whose author is known to one, we would also say of the natural world that it is not altogether going as God would want it to go. We're called upon to begin to see it as we learn God sees it. Rather than giving us specific things or messages, it seems to me that often God is in the business of giving us eyes to see for ourselves, and furthermore what He finally does is give *us*. I mean, He does give us things we need in terms of perseverance, insight, and so on, but ultimately what He does is give us to the world. We ask, 'God what will you give to me?', but we learn to ask, 'God, what will you give me to?'

20.3 What we can learn

AB: When you were asking your first set of questions, I found myself wanting to talk about our second theme. When you were saying, 'Do we just assume that God exists?' Of course, we do look at some of the arguments that God doesn't exist and we think those aren't very good arguments. We also look at some of the arguments that people have used that God does exist. And I think you put it very nicely when you wrote that bit when you said it's hard to disagree when you like the conclusion, but let's nevertheless have a look at whether this argument really stands up to scrutiny.

One of the things I've found myself thinking about, just at the moment, I'd love to know what you think about it, is what we can learn from the very fact that the universe and the world can be understood. What do we learn from the comprehensibility of the universe? Now you have to be a bit careful here because you could say, 'The fact that we can understand the universe is the thermostat on the human intellectual endeavour.' That is to say, we pushed it as far as we can and the thermostat enables us to make progress at such and such a rate but no faster in understanding the world. I'm aware of that. But, if you go past that for a moment, people often talk about the fine tuning of the constants of physics and so on for the universe to support life. I'm thinking of the fine tuning of the comprehensibility of it. In other words, what kind of universe permits a subset to understand the whole, in some measure?

What does it have to be like for that to happen? Is there a sort of fine tuning? In other words, if the world were much, much easier to understand, you could all just understand it in five minutes and that would be the end of the matter. Or if it were much, much harder to understand we'd never understand it at all. So, we couldn't make any progress. It's fine-tuned between those two so that it is comprehensible. And insofar as there is something in that, what does that tell us?

HH: I think the thermostat objection or worry is well placed and I think it should encourage us to be modest, as we don't have an argument which can compel anyone who is resistant to believing these things. But nonetheless when I'm not trying to convince somebody who's sceptical, then I'm just thinking, 'What do I really believe?' I do find it quite amazing. Just in the same way people point out that in some sense it seems somewhat strange that in the universe we human beings are in the middle, there's stuff much, much, much smaller than us, and there's stuff much, much, much larger than us. And here we are medium-sized. In the same way, when you said about understanding, I suspect that in the mid- to late nineteenth century physicists were pretty proud of themselves and what they understood. But we're much more modest now. We realize that we have some great understanding of little localized areas, but as soon as we try to put all these localized areas together, we seem to really run into trouble. And honestly for me that fits extremely well with my theological beliefs because I believe on the one hand, that we were created in the image of God, and on the other that we are finite and fallen. That should immediately put a check on our thought that we are somehow going to achieve this god-like knowledge of the physical world. It actually fits quite well into the actual limitations we have of knowledge, and I think especially coming back to the chapters about quantum physics, I really agree with the spirit of those, as I think if nothing else what quantum physics teaches us is a very heavy dose of modesty in our understanding of physical reality. And that it is thicker, richer, and deeper than people would have expected in the past. For me it actually fits well not with the kind of theism we're criticizing, of the jolly grandfather in the sky, but it fits well with the kind of theism of the hidden God who requires commitment to be known. I do think there's an argument from the intelligibility of the universe.

AB: Perhaps the argument is not so much that God exists, but more: what does this tell us about God?

AS: That's always the question. I would put it back, turn it around. I would not say: does God exist, does God not exist? I would say more like: God is that which calls upon my allegiance, so let's find out who is making that call upon us. I guess the atheist or agnostic wants me to remember that you can't just say 'who'; you can't use such personal language without justification. And I would think, fair enough, that's right. I think that using personal language is a big step. But this much I think I can know: if I want to find out whether the personal language is appropriate, I'm utterly caught in the fact that I can only find that out by taking a personal approach.

HH: Absolutely, I really love the part where we say: wouldn't it be insulting to a person as they sat in a room and we debated whether they exist.

AS: Yes, I hate that phrase: does God exist? It's so pompous.

HH: In a way, any time you are interacting even with a presumed person, you have to approach it differently than you would if it were this purely objective scientific enquiry. Because that is just not the way you treat a person. You don't just say this is completely up for grabs whether you're a person or not.

AS: That's why the question we just alluded to, which I hate, is the wrong question and we tried in our discussion to offer to ourselves and anyone listening in: what other questions are perhaps more helpful? I'll add a couple more. A helpful question might be: How was Bartimaeus healed of his blindness? Did that happen? That to me is an interesting question. And it's not enough to say, 'Of course not.' You have to read a bit further and think it through and, of course, that leads on to other bigger ones such as, 'Is it the case that something tremendously creative and unique happened to Jesus after he died?' because that's a really big question. But at least that is the right sort of question to ask. The other one I would add would be: 'How does it actually pan out in practice, in how you live? What sort of a person does it cause you to become?' That's a good question.

HH: I think that when honestly engaging in intellectual speculation, it's very tempting, especially amongst philosophers, to not remember that we need to return to that question. There is lots to be discussed on the intellectual level but especially important with regard to Jesus is the attitude one has toward what happened after Jesus' death, and a really important thing is 'And so then how should I live in regard to my attitude towards that? What should I ask of myself?' So, it is not purely speculative. We're not merely just doing metaphysics; we're asking, you know, Jesus set an example and we're going to give you some evidence of the example you should follow. I think that is the response that's demanded of us: that this example is what we should set as our paradigm. In some ways, it's pretty daunting. It's not a guide to an easy life.

AS: And the standard is out there now, whether we like it or not!

21

It keeps us seeking

We will now survey what has been accomplished in this book. We will do this by looking back over the four themes that were introduced in Chapter 1. We shall view them in reverse order.

The fourth theme was that we are allowed to open up the window that the natural world offers us. It is a great pity when children are introduced to the study of literature in such a way that it lessens or deadens their receptivity to the beauty and poetry that literature offers. In a similar way, when we approach the natural world we wish to do our science as diligently as we can, but we don't want to close down our receptivity to the possibility that the natural world is also a work of art, or a means by which the human spirit can be enlarged and invited on a journey which is more expansive even than the appreciation of beauty. It is a journey in which we encounter the notion of justice, for example, and the possibility of modes of relation between persons which transcend even that.

In Chapter 5 we suggested that the ability to make choices and be held accountable was at the heart of 'what it means to be me'. We are allowed to entertain that possibility seriously. In Chapters 7 and 8 we noted that the basic nature of the quantum physical world is elusive and not agreed upon. In Chapter 10 we noted that even such an apparently simple thing as empty space can show remarkable subtlety. In Chapters 12 and 13 we found that evolutionary biology does not diminish at all the validity of moral philosophy, nor the notion that humans are responsible agents called to a high purpose. In all these ways we can receive a sort of 'permission to explore'. It is not a permission to make up fantasies, nor to do second-rate intellectual work, but it is a permission to welcome the whole of human nature. Humans have in them a capacity called 'spiritual' which is hard to put into words, but it is real enough, and people are mistreated if this capacity is suppressed. By this capacity, we are allowed to affirm the sense of 'don't stop here' which the physical world presents to us.

But this is not a matter of sure knowledge. Our third theme is that uncertainty is OK. The reader may indeed feel that, at every stage in the book, we have written in a roundabout way and not given pithy solutions and definite 'position statements'. But uncertainty does not prevent progress. It can even be the engine of progress. Often our opinion is, of some way of looking at things, 'this gives some help, but it does not solve all the problems'. That was on show especially in Chapter 15 when we saw how people in previous times have wrestled with religious commitments and scientific knowledge, and in Chapters 16 and 19 on miracles and the Bible. It was also on show when we made some remarks on divine action in Chapter 8 and on the pain of animals in Chapter 13. And just as science lives with uncertainty, and goes forward none the less, so also in our life more generally we can do the same. In Chapter 8 we explained that quantum physicists have no universally agreed view on what quantum 'stuff' is like 'at bottom', but this has not prevented them from greatly extending their understanding and looking forward in eager expectation of wider knowledge to come.

Sometimes we see or hear wry remarks about 'the certainties of faith' or about 'religious conviction' coming from our agnostic friends and colleagues. Such people are wary of the appearance of closure that some kinds of religious attitudes present, and they are especially wary of the coercive or repressive power that religion exerts when it goes wrong. We share that wariness; religion does turn bad when it loses touch with uncertainty. When Jesus challenged people about the nature of their faith, it was not about certainty or even strength of conviction; it was about willingness and resolve. What he wanted was for people to be less fearful, and more willing to try things. He did not talk the language of 'strong faith'. He said that faith is like a seed: just a little is needed. What he did object to was the complete absence of faith, because, after all, faith is not certainty, it is willingness to explore.

Our second theme was the high bar on the quality of argument. If our arguments have not met a high standard, then that is a defect of the book, but in any case we have aimed high. This was on view at length in Chapter 11 on the argument from design, and in Chapter 15 on tension between science and religion, and in Chapters 16 and 17 on miracles, but it was equally important in other parts of the book. When considering quantum physics we tried to state but not overstate what it may have to offer to thinking about the nature of human identity (the soul) and divine action. When considering evolutionary biology we tried to be clear about what is and is not deducible from the natural history of the development of life on Earth, including human life.

In Chapter 19 concerning the Bible we aimed to pay attention to the original language and the situation of the community from which any given text has come. It is right for people to be offered a translation of the Bible into their own native language in the modern day, but words and concepts do not always travel neatly between languages or between historical periods; when difficulties arise, one must enter into an intellectual task which is like a dialogue with the original authors, with a view to finding out what questions they were addressing. In Chapter 10 on general relativity we showed that even in the case of mathematics and physics, where one would think we could get everything clear, it requires much care to figure out what we are really asking and what the answers are telling us. In philosophy, the starting point is often to clarify the language and terms of discussion. Much of philosophical work consists in elucidating what kind of territory one is in. The failure to appreciate this results in very odd mistakes, like the mistake of giving a pair of crampons and an ice axe to someone whose aim is to swim the English Channel.

And now we return to our first theme: God is a being to be known, not a hypothesis to be tested. How can one write about this? Everything we write about God is liable to be more or less impertinent. It is like talking about someone who is there with us, but ignoring them or referring to them as if they were not there. But we can only do, in the pages of a book, what can be done in the pages of a book. Having accepted that restriction, we have tried not to abuse God by treating God like a convenience or an inconvenience, or treating that One like a useful idea or a useless idea, or treating that One like an entity or a non-entity. We can only try to remind ourselves, and our readers, at every turn, that in the end it is the attitude that we adopt that is the most defining aspect of all our attempts to speak of the widest and deepest context of our being. How lovely to be able to express a sense of gratitude, and to be able to acknowledge straightforwardly a sense of our own moral failure and desire for change! How fulfilling to receive the kind of affirmation that assures us we can learn not just to do better but also to be better, to gain better attitudes and assumptions! If that is the sort of thing that we are finding, then we will wish to go further. Since that is the sort of thing we are finding, it keeps us seeking.

22

Thanks

I can no other answer make but thanks,
And thanks; and ever thanks; and oft good turns
Are shuffled off with such uncurrent pay:[1]

In Chapter 5 we mentioned the convention of authors in acknowledging friends and colleagues for their helpful advice and input but taking responsibility for the deficiencies. Now is the time for us to do both. The deficiencies of the book will by now be all too apparent to the reader. We hope that they will tell us what they perceive those to be, and we hope that they will be gentle as they do so. It is with a certain amount of trepidation that one writes openly about things that matter when one is still seeking. But that is the point of the book: to invite others to join us in the quest.

There are many who have inspired us on this quest since long before we (the authors) met each other. It sometimes seems that conscious memory of what you learnt from your parents goes through a minimum at some stage in life, and then starts to climb again through episodic memories of early childhood, and therefore of parental influence. At least that was our experience in writing this book. So we start by thanking our parents, and by extension their parents, and so on back up our family trees. We are also indebted to our teachers, our church leaders, and our classmates and friends at school and beyond. Most recently, Diana Briggs, Keller March, and Emma Steane have given lived examples of wisdom and love, and been generous in their support to us, which enabled us to write. To all we express our thanks. Without them we would not be who we are, and this book would be less than what it is.

There are others who have contributed more directly to this book. Nikki Macmichael has, by her brilliant organization, helped us to get time together to think and write, and also to convene a conference in

[1] *Twelfth Night*, Act III, Scene 3.

Oxford at which the participants gave feedback on a draft. AB's house group at St Andrew's Church, Oxford, spent a term reading an earlier draft and giving invaluable suggestions for improvements; Julia Saunders gave of her professional expertise as a writer and Steve Bagnall compiled a set of discussion questions which will be available on the Web. Colin Humphreys, Pete Jordan, Tom Wright, Brian Heap, Denis Alexander, and Simon Conway Morris each gave scholarly advice on specific chapters. Charles Lauder did more than professional jobs in the development and copy edits. Sonke Adlung and Ania Wronski have given all the support, and more, that any author could wish for in a publisher.

Roger Wagner generously allowed us the use of his painting *Writing in the Dust*. He also gave helpful observations on the whole book, and suggested that we add a further transcript of a discussion, which became Chapters 2 and 20. The photograph in the frontspiece of the three of us was taken by Deborah Lisburne, who also helped to organize the conference, and the title page of Galileo's *A Dialogue Concerning the Two Chief World Systems* is from The Art Archive, Alamy.

This book was made possible through the support of a grant from Templeton World Charity Foundation. The opinions expressed in this publication are those of the authors and do not necessarily reflect the views of Templeton World Charity Foundation. We affirm that in compliance with the contract, but we like to think that Sir John Templeton, with his motto 'how little we know, how eager to learn', would warm to the attitude of *It Keeps Me Seeking*.

At the end of Chapter 4, we write that in the case of a larger piece of work that has more of a sense of personal property, then, we think, an expression of thanks to God, done with due recognition of the multicultural context, and with a light touch, should be allowed. We follow our own advice, and end with humble thanks to God, in whom we live and move and have our being.

Index

Figures and notes are indicated by an italic *f* and *n* following the page number. The number following *n* indicates the footnote number if there is more than one footnote on the page.